50 Thin

you can do with An

by Mike Bedford, G4AEE

Radio Society of Great Britain

Published by the Radio Society of Great Britain, 3 Abbey Court, Fraser Road, Priory Business Park, Bedford MK44 3WH. Tel: 01234 832700. Web: www.rsgb.org

First published 2024

© Radio Society of Great Britain, 2024. All rights reserved. No part of this publication may be reproduced, stored in a retrieval system, or transmitted, in any form or by any means, electronic, mechanical, photocopying, recording or otherwise, without the prior written permission of the Radio Society of Great Britain.

ISBN: 9871 9139 9555 3

Cover design: Kevin Williams, M6CYB

Production: Mark Allgar, M1MPA

Typography and design: Mark Pressland

Printed in Great Britain by Short Run Press Ltd. of Exeter, Devon

Publisher's Note:
The opinions expressed in this book are those of the author(s) and are not necessarily those of the Radio Society of Great Britain. Whilst the information presented is believed to be correct, the publishers and their agents cannot accept responsibility for consequences arising from any inaccuracies or omissions.

Any amendments or updates to this book can be found at: www.rsgb.org/booksextra

Contents

	Contents	*iii*
	Introduction	*iv*
1	Operation	1
2	Construction	47
3	Outdoor Pursuits	81
4	A Personal Perspective	103
5	Beyond the Everyday	117
6	Next Steps	136

Introduction

Are you a radio amateur, considering getting back into the hobby after a break for career or family reasons? Or are you one of the growing band of electronics enthusiasts, inspired by the maker revolution, looking for new avenues to exercise your passion? If so, we have news for you – there's a lot you can do with amateur radio. In fact, if you take the title of this book at face value, there are fifty things you can do with amateur radio. Actually there are a lot more.

If you already hold an amateur radio licence, don't assume that you already know about what you can do with amateur radio. After all, it's constantly evolving. So, quite apart from us reminding you of what first attracted you to amateur radio, we also aim to show you some of the exciting new things you can do today.

Alternatively, if you're an electronics enthusiast, we have some news for you too. Perhaps you're well informed about interfacing real-world hardware to single board computers. But that doesn't mean, there's not a lot of inspiring stuff to learn about using electronics for radio communication. Indeed, there are lots of new and enthralling things to do with radio technology.

All of this might sound convincing, at least superficially, but it's not uncommon to hear the sentiment that amateur radio's outdated. People say "why should I need amateur radio when I can contact anyone in the world with a smartphone?" But that's not comparing like with like. Talking to someone on a phone is a social or business requirement, while amateur radio is a pursuit for those who have an enquiring mind, and that changes everything. You probably wouldn't be surprised to pick up a copy of an electronics or maker magazine and discover an article telling you how to build a digital clock or a robot. Sure you can buy them, but that offers no avenue for your technical and creative skills. In just the same way, building and using an amateur radio transceiver teaches you a lot more about radio frequency engineering than dialling a number on your phone. We'd have to suggest it's a lot more exciting too.

So, what are you waiting for? We invite you to embark on your journey of discovery through the *50 Things you can do with Amateur Radio*, employing your technical skills in challenging and exciting new ways.

Mike Bedford, G4AEE
June 2024

1

Operation

Introduction

Amateur radio is thought of as a technical hobby, and rightly so, but we can't really start our exposé of 50 Things you can do with Amateur Radio with anything other than operation. After all, having a radio station and never using it would be like having a car that you clean every week, and maintain regularly, but which you never drive. Of course, this would involve honing your skills in auto mechanics, and taking satisfaction in your work, but it certainly isn't making the most of the car-owning experience.

This isn't to say, though, that the technical aspects of amateur radio are just a means to an end. After all, some radio amateurs use commercial equipment exclusively, but this doesn't mean that they can ignore the technical aspects of the hobby. Indeed, effective operation involves an appreciation of the technology involved, and of the underlying science behind radio propagation, and so much more. So, you'll learn a lot just by operating your radio station. Certainly, some aspects of amateur radio operation don't require you to exercise your intellect to any great extent, but this just illustrates the diversity of the experiences on offer.

We're starting our investigation of operation by making the somewhat surprising assertion that you can start without needing a licence. While this wouldn't allow you to transmit, you'd learn a lot just by listening and this

50 Things for Radio Amateurs

would probably motivate you to get a licence. Then, to counter the argument that amateur radio isn't needed in a world of cellular communication, we'll explain one important difference. The phone in your pocket rarely transmits a radio signal more than a few kilometres, irrespective of who you're talking to and where they are. But an amateur radio station can genuinely transmit to the other side of the world!

Our next port of call concerns the radio spectrum. We'll start by investigating this important segment of the electromagnetic spectrum before moving on to look at three areas of that spectrum, which differ so much in what they offer. In particular, we examine the shortwave, VLF and microwave regions, investigating what each offers and the challenges they present.

Next up, we think about how you can use amateur-built repeaters to extend your range, perhaps when you're using a low-power VHF radio in your car. For this you'd use a terrestrial repeater, but our next topic goes one step further in looking at repeaters that are orbiting the Earth on board amateur radio satellites.

Radio amateurs refer to modes of operation and these are our next theme. Speech communication is surely the most obvious mode, but it's not the only one. It's not for everyone but the first mode we'll look at is Morse code, which is still going strong and can often outperform verbal communication. We then proceed to look at state-of-the-art digital communications. This then leads to our final modes which allow text to be transmitted using a signal that's so weak it can't even be heard.

We conclude our look at operation by considering a miscellany of topics. These range from understanding the scientific processes in the atmosphere that make long-distance communication possible, to the quest to contact every country in the world – all 195 of them.

1: Operating

Thing 1 – Get Started Without a Licence

Amateur radio differs from other freely-available types of radios like CB rigs and PMR 446 walkie-talkies by requiring that the operator is technically competent. This means that radio amateurs are required to hold a licence, and these are only granted if the applicant passes the relevant examination. However, once licensed, radio amateurs are given access to large swathes of the radio spectrum and are permitted to use high power. The upshot of this is that, while PMR 446 and CB users can rarely achieve a range of more than a few kilometres, radio amateurs can communicate with others around the globe.

This might seem to question our suggestion that you could make a start without needing a licence. However, getting started doesn't necessarily mean transmitting, and it's probably true to say that most amateurs took their first steps by listening on the amateur bands. If you're new to the concept of the amateur bands it would be a good idea to take a look at our introduction in Thing 3, but most people start on the shortwave bands – Thing 4. After all, these are the bands that allow worldwide communication and, if you need encouragement to get a licence, what could be better than hearing radio amateurs in New Zealand or Fiji? There's no better way of becoming sufficiently motivated to pass the necessary examination and getting a licence than listening to radio amateurs in such far-flung places.

A Receiving Station

Although there's an alternative method, the most obvious way to start is to get hold of a suitable receiver and connect it to a decent antenna. We provide some guidance on installing antennas in Thing 24. The subject of buying a receiver is important because, if it doesn't pass muster, your first exposure to amateur radio might not show it off in a good light. However, if you plan to obtain a licence, buying an expensive receiver and selling it to buy a transceiver only a few months later might not be cost-effective. In passing, though, if you are committed to getting a licence, don't discount the idea of buying a transceiver for your first steps. Owning one without having a licence isn't illegal as long as you don't use it to transmit.

Prices for radios that include shortwave coverage vary from less than £50 to several thousands of pounds and specifications differ substantially, so choosing one isn't necessarily easy. The cheapest radios are portable devices intended for general use such as listening to broadcast stations, but some of these can be used on the shortwave bands. So

Photo 1.1 Some cheap portable radios offer shortwave coverage but they're not all suitable for receiving amateur stations. So be sure to consult reviews before buying. Photo: James Case.

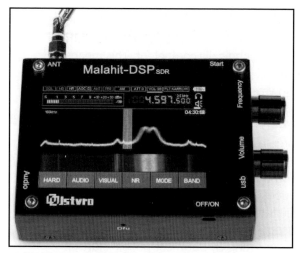

Photo 1.2. Radio receivers intended for radio enthusiasts offer decent performance but cost more than general-purpose portable radios.

they might also be suitable for listening on the amateur bands. However, you should make sure any prospective radio covers the portions of the shortwave spectrum you're interested in (see Thing 4) because some have limited shortwave coverage. Also, check that it supports SSB transmissions. We don't need to go into the details here but without SSB support, you'll be able to receive shortwave broadcast stations, but not most amateur stations on the shortwave bands. These types of radio differ a lot in their performance, with reviews suggesting that some offer very poor performance of amateur radio transmissions. So, be sure to read reviews before choosing one of these radios. Alternatively, you could choose a radio that's intended specifically for use by enthusiasts. These are more expensive, with prices starting at around £100, but they'll almost certainly allow you to hear many more stations. Again, do check reviews first.

The Web SDR Alternative

We now move on to look at the somewhat counter-intuitive suggestion that you can start listening to amateur radio stations even if you don't own a suitable receiver and haven't installed any antennas. This doesn't mean that you won't be using a receiver and antenna, but that you'll be using a remote receiving station set up by someone

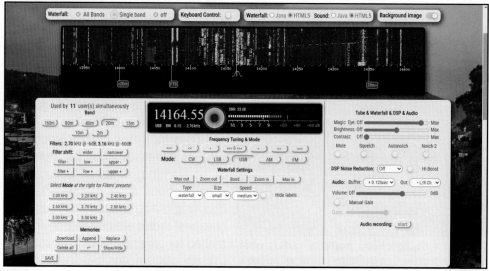

Figure 1.1. You can even get your first experiences of the amateur bands without buying a receiver, thanks to WebSDRs.

1: Operating

else and hosted online. What we're referring to is using a WebSDR, and you can choose one at www.websdr.org. You'll find these receiving stations all around the word, but you might prefer to use one in your home country, or nearby. This will give you the best feel for what you might hear when you set up a station at home. None of these radios will cover all of the amateur bands, after all there are so many of them, although most provide coverage of many, if not all, of the shortwave bands. Even so, your choice might be reduced somewhat, depending on which bands you want to listen to.

Most of the online receivers present a similar on-screen appearance. There will be controls to allow you to choose an amateur band and the type of transmission, which will usually be CW for Morse code, LSB for speech on frequencies up to 10MHz – mostly the 160m (1.8MHz), 80m (3.3MHz) and 40m (7MHz) bands – and USB for speech above 10MHz. An important part of the on-screen representation is a so-called 'waterfall display' which provides a graphical representation of signals across your chosen band. Clicking in the area immediately below this display allows a particular signal to be selected. This allow you to hear any station on your chosen frequency, although commonly, you first have to click on a button labelled 'audio start' or similar. There are a lot more controls, but they're fairly intuitive so you should soon get up to speed.

Keep a Log

Although it's no longer a legal requirement in the UK, radio amateurs often keep a logbook, detailing all of their contacts. It might be a good idea to get into the practice of logging. This brings with it several benefits. For example, looking back over your log will help you to get a better feel for the different characteristics of the various bands. Certainly you can read that long-distance communication is more likely on the 20m (14MHz) or 15m (21MHz) bands than on the 80m (3.5MHz) band, for example, but there's something special about discovering these things yourself.

Your logbook could take the form of a traditional book, but using logging software, some of which is free, offers so many more options. Do read up on the functions on offer but, irrespective of the features of a particular application, there are plenty of related packages which can import your log file. This may allow you to analyse your log in various ways, or even display stations on a map. Alternatively, some logging tools allow you to export your data in a common format, thereby allowing you to carry out your own analyses in a spreadsheet.

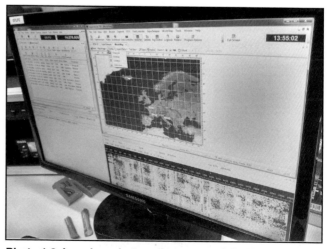

Photo 1.3. Logging what you hear, especially if you use software, will give you a better appreciation of the amateur bands.

Thing 2 – Span the World with no Base Stations

Given that you can easily make contact around the world using a tiny mobile phone, reference to amateur radio allowing you to span the world might sound irrelevant. But we're not comparing like with like. Although the experience of using one seems to suggest otherwise, a smartphone really can't transmit a signal very far. In fact, the signal only needs to reach the nearest base station – or 'mast' as it's called colloquially – which we can think of as a repeater. Commonly, when you use your phone, the base station will be within a couple of kilometres , and it appears that the maximum range is about 40km. What's more, these distances are constantly decreasing as new generations of cellular technology provide higher rates of data transmission by using ever high frequencies. From the base station, the signal will, sooner or later, be transmitted by cables, including submarine cables, for long distance calls. And it'll only become a radio signal again at the far end if it's received by another mobile device. So, any thought of that phone call from London to Sydney being carried by radio is somewhat short of the mark.

With amateur radio all this changes. Some radio amateurs do use repeaters (see Thing 7), but generally only to extend the range of low-power mobile stations or walkie-talkies operating in the VHF or UHF bands. On several of the shortwave bands, however, it's entirely feasible to make contact with a radio amateur on the other side of the globe in a single hop. Actually, the word 'hop' is entirely appropriate here, because of the way shortwave signals make their way around the world. As you can read in Thing 12, shortwave signals bounce, between the Earth's surface and that part of the atmosphere known as the ionosphere. And with several bounces, or hops, they can reach anywhere in the world. When we consider that radio waves usually travel in straight lines, this method of global communication is surely astonishing. Yet it's just one of the amazing things you can do with amateur radio.

Photo 2.1. Amateur radio signals genuinely span the globe whereas mobile phones can only achieve the few kilometres to a base station. Photo: Les Chatfield.

Thing 3 – Discover the Radio Spectrum

Radio waves are a form of electromagnetic radiation so, in that sense, they're similar to light. Where radio differs from light, though, is in its frequency. Radio has a much lower frequency than light, falling directly below that of infrared radiation. Specifically, although these figures are somewhat arbitrary, radio waves are often considered to have a frequency below 3THz. You can see where radio fits into the electromagnetic spectrum in **Figure 3.1**, and you can see the most important parts of the radio spectrum in **Figure 3.2**. Confusingly, there is an overlap between the high-frequency end of the radio spectrum and the low-frequency end of infrared.

In fact, Figure 3.2 doesn't show the entire radio spectrum, although it does include those parts where most amateur radio operation takes place. Each of the boxes represent a radio band, as defined by the ITU (International Telecommunication Union), although these aren't the same as amateur bands, some of which we'll encounter in Things 4, 5 and 6. In addition to the bands shown in Figure 3.2, there are also the ELF (Extremely Low Frequency, SLF (Super Low Frequency) and ULF (Ultra Low Frequency) bands at the bottom end, and THF (Tremendously High Frequency) at the upper end. As you'll notice, each band covers an order of magnitude, so the upper end of each band has a frequency 10 times higher than its lower end. We also show wavelengths, and these decrease as frequencies increase. To illustrate the difference between ITU bands and amateur bands, we've shown several black vertical bars in the MF and HF bands. These represent ranges of frequencies where radio amateurs are allowed to transmit in the so-called 'shortwave region'. These are called 'amateur

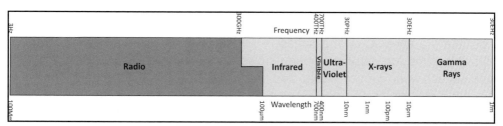

Figure 3.1. Radio covers the bottom part of the electromagnetic spectrum, below infrared, visible light, ultraviolet, X-rays and gamma rays.

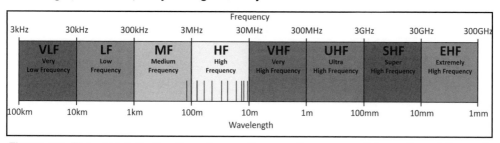

Figure 3.2. The vast majority of amateur radio operation covers this portion of the radio spectrum.

bands', and there are others within the other ITU bands. Some countries also allow amateur radio operation at any frequencies around 9kHz in the VLF band, while others allow any frequency below 9kHz to be used. In addition to VLF, that would also include ULF, SLF and ELF. Also, while there are no amateur bands at THF worldwide, amateur operation in this ITU band is allowed by special permission in some counties.

The radio spectrum, even if we only include the portion shown in Figure 3.2, covers eight orders of magnitude. This is far more than any other recognised portion of the electromagnetic spectrum. Visible light, for example, covers wavelengths from about 380nm to 750nm, which is the equivalent of frequencies from 400THz to 900THz, which is less than one order of magnitude. To give a better feel for just how much the normal amateur portion of the radio spectrum covers, we should point out that the wavelengths represented by the eight ITU bands ranges from 1mm to 100km. If this huge range of wavelengths and frequencies makes a difference, we could expect some massive differences in a whole manner of things. And the good news, if you think variety is the spice of life, is that wavelengths and frequencies do make a difference, and a big one at that.

Perhaps the difference that will be of most interest to newcomers to amateur radio is the range over which you could transmit a signal. It's hard to make direct comparisons because so many other things will change as you move through those eight orders of magnitude. However, as referred to in Thing 2, it's possible to achieve global coverage in parts of the HF band, whereas ranges are much reduced at VLF, while in the 1mm (284GHz) amateur band in the EHF region, the world record is just 114km. Higher frequencies come into their own, though, when we consider the types of signals that can be exchanged. In many of the amateur bands, and certainly those at HF, communication is possible via speech, or messages in Morse code or more modern modes of data transmission. On lower frequencies, the available bandwidth is reduced and speech is no longer viable, or not allowed. On the other side of the coin, amateur TV signals (Thing 43) can be transmitted on the 70cm (430MHz) band and, at higher frequencies, high-definition TV is possible.

Technical differences also abound as we move throughout the radio spectrum but we're not going to get embroiled in that here. However, some of these will come to the fore as we think about transmitting on the shortwave (HF), VLF and microwave (top end of UHF and above) bands in the next three 'things'.

1: Operating

Thing 4 – Transmit on Shortwave

As we start our investigation of transmitting in various parts of the radio spectrum we come to the shortwave band. We're starting here because, in many ways, these bands can be thought of as archetypical amateur radio territory. We say this in recognition of the fact that it's here that worldwide coverage is possible. You won't find anything called 'shortwave' in Figure 3.2, though, because it's not an official designation, even though it's a commonly-used term. The shortwave region of the radio spectrum can be considered as approximately the same as the HF (High Frequency) segment, although it's often extended to include some or all of the MF (Medium Frequency) segment. Here we're looking at the amateur bands from 160m (1.8MHz) to 10m (28MHz). And in terms of equipment, most commercial transceivers cover all the shortwave bands, and probably a few VHF bands too.

The use of shortwave might be considered somewhat outdated, but it continues to be utilised, and not just by radio amateurs. Despite the proliferation of direct satellite broadcasting and internet radio, the shortwave bands are still used for broadcasting over large areas, especially where internet infrastructure is limited. It also continues to be used for long-range marine communication, and for military applications, presumably because it's more difficult to jam than satellite transmissions. In the realm of amateur radio, though, it's mostly used because it poses interesting challenges, and it's fascinating that it can provide global coverage using cheap, low-power equipment. What's more, even though shortwave covers a small portion of entire radio spectrum, different shortwave bands have very different characteristics. Here we're going to provide an overview of the amateur shortwave bands by looking at a representative cross-section of the ten shortwave bands. We're not going to explain why they differ so much, though. However, we look at the subject of radio propagation in Thing 12, and this provides some

Photo 4.1. A single commercial transceiver can provide coverage of all the shortwave bands. Photo: Dave Parker.

9

insight into why these bands behave in such different ways.

Top Band

The first band on our whistle-stop tour of the shortwave bands is the 160m band (1.8MHz), commonly called 'Top Band' by radio amateurs. It's actually the lowest frequency shortwave band, but this nickname is possibly a throwback to the days when amateurs tended to think in terms of wavelengths.

Given our suggestion that the shortwave bands can provide worldwide coverage, your first foray onto the 160m (1.8MHz) band might be something of an anticlimax. During the day you might find it especially unimpressive, with most stations you hear or contact being within 150km or so. First impressions can be deceptive, though, and that would be the case here. Indeed, as day turns into night you'd find that communication becomes possible over a greater range, 1,000km or more being not at all unusual. And it gets better. Although it would be the exception rather than the rule, at times around sunrise and sunset, especially during winter months, much longer distances are achievable. Indeed, some amateurs have contacted more than 100 countries on top band. We do have to point out, though, that DX, which is amateur radio parlance for 'long distance', isn't necessarily easy on 160m, and perseverance is certainly needed.

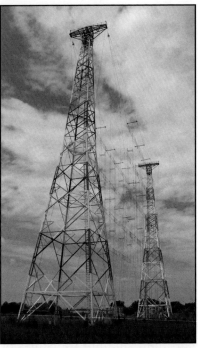

Photo 4.2. Broadcast stations like this one in Australia continue to prove that shortwave radio offers several benefits, something that's not lost on radio amateurs.

Part of the reason that DX is difficult on 160m (1.8MHz) is because of the band's unique propagation characteristics. However, that's not all. As you can read in Thing 24, the common dipole antenna is half a wavelength long, and a popular vertical antenna is a quarter or a wavelength long. On top band, these becomes 76m and 38m respectively, lengths that often aren't feasible. There are other types of antenna that are smaller but, as sizes drop significantly below a quarter of a wavelength, performance suffers.

If you're new to amateur radio, you might be wondering why amateurs try to make long-distance contacts on top band, despite the difficulty, when it would be so much easier on other shortwave bands. We're not going to answer that question, but we will pose a rhetorical question, which might shed some light on 160m band DX attempts. Why did Eliud Kipchoge spend so much time in training for the 2022 Berlin Marathon, at which he took a world record time of 2:01:09, when he could have driven the Marathon distance in less than half an hour?

80m and 40m Bands
As the next higher frequency band to 160m (1.8MHz), we might expect the 80m band (3.5MHz) to behave in a similar way, and in some ways it does. It's also true to say that antenna constraints also apply to those without much outdoor space. In turn, this will compromise performance compared to, for example, the 20m (14MHz) and 10m (21MHz) bands. So what could you expect of the 80m (3.5MHz) band?

Probably the main thing that differentiates 80m (3.5MHz) from 160m (1.8MHz) is that normal ranges increase, as does the ease of making DX contacts. In fact, this is a trend that continues as we move upwards in frequency although, as we'll see, these trends don't continue indefinitely. In particular, daytime ranges are greater, so British radio amateurs can generally make contacts across the UK and perhaps into the closest parts of mainland Europe during daylight hours.

Similar trends continue in the 40m (7MHz) band, so daytime contacts of over 1,500km are possible and, at night, DX operation is easier than on 160m (1.8MHz) or 80m (3.5MHz). This is due, in part, to antenna constraints being relaxed, indeed high-gain rotatable Yagi antennas (see Thing 24) are just about possible, even though they are huge and expensive.

20m and 10m Bands
When we reach the 20m (14MHz) and 10m (28MHz) bands, some but not all of the trends we've seen so far continue. In particular, the ease of making DX contacts, even with the station on the opposite side of the globe, becomes substantially easier. However, there are several nuances.

For a start, despite the potential for DX, the possibility of local communication during hours of daylight no longer applies. In fact, except for very local coverage, perhaps up to 40km, there's a dead-zone before more distant communication becomes possible. Next up, unlike the lower frequency bands, long distant communication is no longer limited largely to night time. And third, DX on 20m (14MHz) is improved when there are lots of sunspots, while at 10m (28MHz), a high sunspot count is almost essential. As we'll see in Thing 12, sunspots fluctuate over an 11-year cycle and the peak is expected in a year or so (at the time of writing in early 2024). The one area in which our trends do continue, though, is that high gain antennas become much more feasible. At 20m (14MHz), the serious DXer can, indeed, aspire to install a three-element Yagi antenna. Indeed, tri-band Yagi antennas, which also cover the 15m (21MHz) and 10m (28MHz) bands, are justly popular.

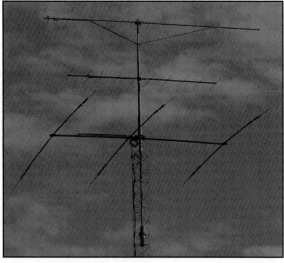

Photo 4.3. High-gain rotatable antennas like the lower of the three antennas on this tower, become practical at 20m (14MHz) and above.

Thing 5 – Transmit at VLF

Our next segment of the radio spectrum that we're looking at is the ITU band designated as VLF (Very Low Frequency) which covers 3kHz to 30kHz. We've chosen to feature this band because it's different, in so many ways, to the shortwave bands we saw in Thing 4. We do have to admit, though, that there isn't a VLF amateur allocation in all countries, but that doesn't mean there's no amateur operation here. In the UK, for example, frequencies in the range of 8.7-9.1kHz can be used by radio amateurs, but only with special permission, in the form of a Notice of Variation to their licences. VLF allocations are different in other countries. However, in some countries there are no restrictions at all on transmitting on frequencies below 9kHz.

Performance

When amateurs first started using frequencies around 9kHz it was dubbed 'the dreamers' band'. Despite the implication that achieving long distances would be the stuff of dreams, these dreams have actually come true in recent years. For while some of the first tests achieved ranges measured in single figures of kilometres, these figures have progressed in leaps and bounds. Indeed sub-9kHz signals have now crossed the Atlantic. However, things are not all that they seem. Covering significant distances can only be achieved using very slow digital or Morse code signals. In fact, transmission times measured in hours or days have been needed to exchange just a few characters. What's more, it's necessary to ensure that the transmitting and receiving stations are synchronised, commonly by the use of GPS signals. Needless to say, rarely will someone just stumble across one of these transmissions so, instead, contacts are nearly always achieved by scheduling a test in advance.

If you thought that frequencies near the bottom of the VLF bad were pretty much as low as you can go, think again. Amateurs in countries where regulation of frequencies below 9kHz is more relaxed, and especially in Germany, amateurs have gone lower, much lower. We note, for example, that radio amateur Stefan Schäfer, DK7FC, has been experimenting on 22.97Hz, yes, that's Hz, not kHz. This is no longer in the VLF band, nor is it in the UHF band, which spans 300Hz to 3kHz, nor the SLF band that covers 30Hz to 300Hz. Instead, it's in the ELF band, which surely is about as low as you can go. The range achieved by Stefan wasn't a lot, but he did manage 3.5km. To be fair, though, he did estimate the ERP (effective radiated power) at just 600aW,

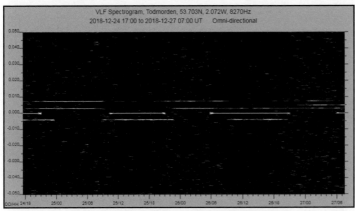

Figure 5.1. This waterfall display shows five VLF signals, from Germany, Italy and Canada, received in Todmorden, Lancashire, UK. The shortest signals took about 12 hours to receive. Image: Paul Nicholson.

that's less than a billionth of a microwatt.

Antennas
It's commonly considered that efficient antennas should be at least a quarter of a wavelength long, and transmitting antennas for most shortwave frequencies and above are rarely shorter. At 9kHz, though, that quarter of a wavelength would be 8km. While antennas of this sort of size and above have been used by the US Navy for submarine communication, this doesn't even come close to being a practical proposition for radio amateurs. At Stefan Schäfer's 22.7Hz frequency, a quarter-wave antenna would be more than 3,000km long.

A common technique in the HF bands, when it becomes necessary to use an antenna that's appreciably shorter than a wavelength, say for mobile operation (Thing 25), is to use a so-called loaded antenna. They are much less efficient than a quarter-wave vertical antenna, but adding a loading coil to the base of a short whip allows it to be matched to the transmitter. For the amateur shortwave bands, that loading coil might have 50 turns on a former measuring 250mm long and 50mm in diameter. At VLF this wouldn't scratch the surface. If you want to get a feel for what's needed, take a look at **Photo 5.2**.

Photo 5.1. At VLF and below, efficient transmitting antennas are huge. This station at Clam Lake, Wisconsin, used an antenna 22.5km long.

Getting Practical
So is VLF experimentation largely a pointless challenge that's undertaken "because it's there" to quote George Leigh Mallory's speaking of his upcoming quest to conquer Everest? In part, yes, but that's only part of the story. It can be argued that no pure research is ever wasted, and a lot of new ground has been broken in taming this most unforgiving portion of the radio spectrum. More down to earth, though, and quite literally, as you can read in Thing 36, radio amateurs have used VLF and LF radio to provide communication for cave rescue teams.

Photo 5.2. This loading coil, made by German amateur Bernd Mehlis, DL3JMM gives some idea of the lengths radio amateurs have gone to.

Thing 6 – Transmit on Microwaves

From the very bottom end of the radio spectrum, we're now moving to the very top. Often referred to as microwave bands, we're thinking of the amateur bands that have a frequency in excess of 1GHz. This includes part of the ITU UHF (ultra high frequency) band and the entirety of the SHF (super high frequency), EHF (extra high frequency) and THF (tremendously high frequency) bands. And in terms of the amateur bands, there are no fewer than twelve microwave bands which range from the 23cm (1.2GHz) to the 1mm (248GHz) band. Access to yet higher frequencies is available via a notice of variation to an amateur radio licence in the UK.

Our motivation in presenting the microwave bands as one of our "things" is quite different to our reasons for looking at the other end of the radio spectrum, namely the VLF band. While recognising that there's one important real-world application in the form of cave communication (see Thing 36), amateur VLF experimentation is mostly done because of the challenges it presents. But while practical applications of VLF are limited, the same cannot be said of microwave communication. In particular, these frequencies are used for satellite communication, for mobile phone networks and in any wireless-enabled equipment using various communication standards including Wi-Fi, Bluetooth and ZigBee. So, in experimenting with microwave frequencies, radio amateurs are, potentially, furthering state-of-the-art technologies.

Distance Records

Of course, this doesn't mean that radio amateurs don't engage in attempting to break distance records in the microwave bands. After all, it's surely part of the amateur radio mindset to take part in this sort of activity. Unlike the case at VLF and below, where distance decreases as the frequency decreases, at the top end of the radio spectrum ranges decrease with increasing frequency. So it's probably appropriate to say something about a record on the (1mm) 241GHz band. When amateurs first started on this band

Photo 6.1. This tripod-mounted equipment was used by G8ACE and G8KQW to achieve a new distance record on the 2mm (134GHz) band.

in the UK, only a few years ago, they made contact over just 30m. Today, though, the word record for a terrestrial contact, as opposed to one involving communication via a satellite, is 114km.

Challenges and Solutions

As with VLF communication, commercial equipment isn't available for most microwave amateur bands, so radio amateurs have to build their own gear. At VLF, the main challenge is in building the antenna, while in the microwave bands, the transmitter and receiver design pose serious difficulties. In particular, obtaining a high transmitter power is seriously tricky, and the output is typically measured in low numbers of microwatts in the highest frequency bands such as 1mm (241GHz). This is relieved, to a degree, when we consider antennas. At VLF, despite being able to output tens or hundreds of watts from the transmitter, the effective radiated power is often a miniscule fraction of a watt, because of the inefficiency of the antenna. At the higher microwave frequencies the converse is often true, because the wavelength is so short that high-gain antennas are eminently feasible. Commonly parabolic dishes, these antennas could provide a gain of 50dB, which would increase the transmitted power output to give an effective radiated power of 100,000 times greater. Things aren't quite that simple though. This sort of gain is achieved as a result of these antennas' directionality. In other words, the width of the transmitted beam is very narrow. Aligning the antennas at each end of the path can be extremely difficult, therefore, and the smallest of errors can make all the difference between success and failure. You might want to ponder, therefore, on how that could possibly have been achieved on 241GHz over the 114km path of that world record breaking contact.

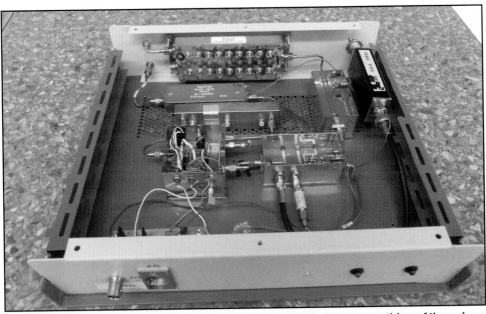

Photo 6.2. This transverter, built by John Worsnop, G4BAO shows something of the unique challenges of building equipment for the microwave bands.

Thing 7 – Use Repeaters

If you're using a car-based mobile station on a VHF or UHF band, or if you're using a handheld radio, your range might be severely limited. This is especially true if you're in a built-up area surrounded by tall buildings. If you're in a rural area, your range could be improved significantly, but you still might not be able talk to anyone because such areas tend to be sparsely populated. In both of these cases, amateur-operated repeaters provide a solution. These repeaters are used to simultaneously receive and re-transmit signals on a slightly different frequency. This invariably improves the range for mobile or handheld users because the repeater will use more power, and is often located in a favourable location such as a hilltop. Most repeaters operate on FM – which is almost universal for mobile or handheld VHF and UHF rigs, and is what we're assuming here – although some repeaters can handle data or TV signals. In addition, most repeaters work on the 2m (144MHz) band, with the 70cm (430MHz) band being the next most common. To make use of a repeater there are several things you need to know. These are concerned with accessing a particular repeater, and general operating principles.

Accessing a Repeater

There are a couple of things you need to know, and program into your transceiver, in order to use a particular repeater. These important pieces of information can be found online, but only if you can identify a repeater you're listening to and which you want to access. Fortunately, most repeaters identify themselves periodically by transmitting their callsign. This will be in either Morse code or speech, depending on the country. Alternatively, if you're going to be travelling away from home, you could search out several repeaters online and programme their frequencies,

Figure 7.1. Repeaters allow two mobile or handheld stations to communicate where a direct link wouldn't be possible.

offsets and CTCSS tones into different preset memories.

First of all, because repeaters transmit and receive at the same time, they have to transmit and receive on different – albeit usually nearby – frequencies. The difference in frequencies is often given as an offset, this being the repeater's input frequency (the frequency you transmit on) minus its output frequency (the frequency you listen on). So, for example a repeater that listens on 145.125MHz and transmits on 145.725 has an offset of -600kHz. Your transceiver can be configured to transmit and receive on different frequencies, but you'll need to look up the offset for the repeater you plan to use.

The next thing you need to know is how to access a repeater. To prevent the repeater being triggered accidentally, or by noise, stations using a repeater need to start their transmission with a sub-audible CTCSS (Continuous Tone Coded Squelch System) tone. Different repeaters use different tone frequencies, so this is something else you'll need to find out for a repeater you intend to use and program it into your transceiver.

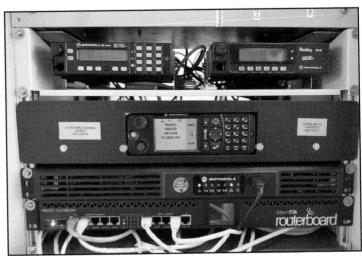

Photo 7.1. Chances are you'll never see a repeater up close, but that doesn't mean they don't offer an important service. Photo: Ptolusque.

Operating Practice

Having set up your transceiver to access a particular repeater, there are a few general things you need to know to access any repeater. First of all, in most case, repeaters are open access. In the UK, though, radio amateurs are expected, but not compelled, to pay a small subscription to help fund the repeater they use most often.

With the transceiver appropriately configured, accessing a repeater is a simple matter of making a transmission when nobody else is accessing it. Beware, though, that many repeaters impose a time limit on any transmission, and will time you out if you exceed this limit. So, find out what the limit is and keep to it but, in general, try to keep your transmission succinct and to the point. Perhaps the only other thing to know is that, if you're engaged in a conversation, repeaters often transmit an audible tone, of perhaps the letter K in Morse code, when a station has ended their transmission. This, therefore, is your prompt to reply. If in doubt about anything, though, spend some time listening to repeaters in general, or specific repeaters in particular, before making your first transmission.

Thing 8 – Use Amateur Satellites

If transmitting via a terrestrial repeater is good, then surely using a repeater on board a satellite hundreds or even thousands of kilometres above your head is even more amazing. But despite being undeniably high-tech, it's entirely possible, and not even as difficult as some endeavours in amateur radio. Let's start with some background information.

Satellite Basics

The original amateur radio satellite, OSCAR 1, was lunched in 1961, only four years after the launch of Sputnik 1, the first ever artificial Earth satellite. It didn't carry a repeater, though, instead it carried a simple beacon which transmitted a Morse signal on the 2m band. OSCAR 3 went into operation in 1965 and became the first amateur-built satellite to carry a repeater. Satellites don't work forever, though, indeed this particular satellite operated for only 16 days.

But where OSCAR 3 led, other satellites followed, and today there are around 20 satellites serving the amateur community. There are two main types of amateur satellite, those carrying an FM repeater, and those that have a transponder on board. Those with an FM repeater operate in much the same way as a terrestrial repeater, but with one important difference. Instead of transmitting and receiving on different frequencies within the same band, their transmit and receive frequencies are in different bands. The most common combination is an uplink in the 70cm (430MHz) band and a downlink in the 2m (144MHz) band. Transponder satellites differ in that instead of transmitting and receiving single frequencies, they transmit and receive a band of frequencies. This allows several radio amateurs to use the satellite at the same time, and SSB and CW transmissions are commonly used. Some transponder repeaters are designed specifically for data. Like FM repeater satellites, they operate in a cross-band fashion, although the 2m (144MHz) /10m (28MHz) combination is a common alternative to the use of the 2m (144MHz) and 70cm (430MHz) bands.

Nearly all amateur satellites are of the LEO type. Standing for 'Low Earth Orbit', these satellites orbit the Earth several times per day, and are within sight of any location on Earth two or three times each day. They are at an altitude of a few hundred kilome-

Photo 8.1. There's surely no more impressive illustration of amateur radio than this satellite being launched atop a Falcon 9 rocket.

1: Operating

tres, although when they're near the horizon, they'll be at a range of over 1,000km. There is also one geostationary amateur satellite, namely Es'hail-2, otherwise known as OSCAR-100. Geostationary satellites are at an altitude of 35,785km, which means that have an orbital time of 24 hours. Accordingly, they appear to be at a stationary position above the Earth, and provide coverage to almost half the globe. Es'hail-2 is within range of all of Europe and Africa, much of Asia, and part of South America.

The other thing to mention is that the cost of designing, building and launching satellites has decreased significantly over the decades. This, in turn, can only be good news for radio amateurs. Of particular interest are CubeSats. Officially designated as nanosatellites, a CubeSat has a standard form factor, measuring 100mm x 100mm x 100mm, and weighing no more than 2kg. Alternatively, larger satellites can comprise several of these basic modules. Already there are several CubeSats carrying amateur radio payloads, and we can surely expect to see this trend continue.

Satellite Operation

So that's the background, but how can you actually start using these repeaters? The bottom line is that it's nowhere as simple as using a terrestrial repeater so it will be quite a learning exercise. However, we'll provide you with a very brief overview of the techniques and challenges in the hope of whetting your appetite.

First of all let's look at the equipment you'll need. You have to be able to transmit and receive on different bands and, what's more, you need to be able to operate in full-duplex mode. This means that you need to be able to transmit and receive at the same time. Many transceivers are incapable of doing this but, if yours doesn't, you can use two different radios. Next you need antennas for both frequencies, and these need to be reasonably high-gain directional antennas such as Yagis. Tracking a satellite as it moves through the sky with two antennas is unnecessarily difficult so, ideally, you should use a dual-band antenna. These are commercially available or you could build your own.

Next, you need to be able to find a satellite in the sky. Several websites provide

Photo 8.2. This FASTRAC satellite, which was made available to radio amateurs, was designed by students at the University of Texas in Austin.

Photo 8.3. Cubesats represent a new era in satellite design, and their small size and low cost could revolutionise amateur satellites. Photo: Svobodat.

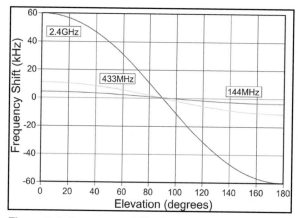

Figure 8.1. Frequency shift, due to the Doppler effect, is a factor that has to be taken into account when tracking a satellite.

constantly-updated lists of all satellites that will be visible from your location and their position in the sky. Also, most satellites have several modes of operation. Specifically, they don't always have their repeater or transponder turned on and, when they do, there are various possible combinations of uplink and downlink frequencies. Again, this is something you'll need to check using real-time web resources. When you find a satellite, or at least when you're pointing your antenna in the correct direction, you need to check that you can access it. The bottom line is that, if you transmit and can't hear your signal being returned, nobody else will hear you either. Being able to check in this way is one of the reasons that you need full-duplex operation.

The final thing to mention concerns Doppler shift. When a satellite is moving towards you in its orbit, its frequency increases, and when it's moving away, its frequency decreases. This is purely a fact of physics, and you'll have noticed the same sort of thing, but in the audio domain, when an emergency vehicle passes you with its siren sounding. The presence of Doppler shift means that you need to track the frequency of a satellite as it travels from one horizon to the other. Fortunately, the degree to which the frequency shifts on the 2m band is minimal, so you only need to track on the 70cm band (430MHz) which, for FM repeaters, is your transmitting frequency. So, when you're transmitting, you need to monitor the signal being returned to you on the 2m band (144MHz) and adjust your 70cm (430MHz) band transmit frequency so it remains intelligible.

The Es'hail-2 geostationary satellite presents its own challenges, mainly because it's much further away so accessing it will require higher gain antennas, commonly parabolic dishes. In addition, you'll be moving up the frequency range compared to most amateur satellites. Es'hail-2's uplink is in the 13cm (2.3GHz) band, while its downlink is in the 3cm (10GHz) band. Picking up the gauntlet and building a station capable of transmitting via this satellite could well be worth the hard work, though. After all, the transmission path, up to Es'hail-2 and back down again is almost two times the Earth's diameter, or a fifth of the distance to the Moon. Even if you're not ready to take up the challenge of all this just yet, you can listen to traffic on the satellite's down link via the WebSDR at https://eshail.batc.org.uk/nb/.

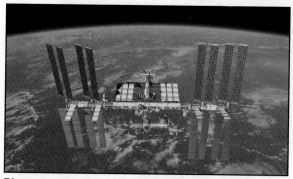

Photo 8.4. A rather special case of an amateur radio satellite is the one that's carried in the International Space Station.

1: Operating

Thing 9 – Make Use of Morse Code

The subject of Morse code might be unexpected if you don't have a background in amateur radio. So we really need to start by addressing the question of why it's still used by radio amateurs. After all, it's no spring chicken, in fact it dates back to around 1837.

Why Morse?

The first reason that Morse code is still used is that it allows some almost unbelievably-simple equipment to be used. Indeed, we discover the Pixie CW transceiver, that used just two transistors and an IC, in Thing 21, and a single transistor CW transmitter in Thing 22. CW stands for 'continuous wave' and is the means by which Morse code is transmitted by radio. We do recognise that Australian radio amateur Peter Parker, VK3YE converted a Pixie to transmit a voice signal without adding any more transistors or ICs. However, this was at the expense of its receive capability, and if you used another Pixie as the receiver, there would be four transistors and two ICs in total. We also have to point out that this modification allowed speech to be transmitted via AM (Amplitude Modulation), which is not nearly as efficient as SSB (single sideband), which is the voice mode that is most commonly-used by radio amateurs. And there's no way an SSB transceiver, or even just a transmitter, could be built without using much more complicated circuitry.

The second reason is that, all other things being equal, CW allows much greater ranges to be covered than speech, or even SSB. Intuitively, this isn't at all surprising. Imagine listening to a weak signal that's partially obliterated by interfering signals. Surely it's far easier to distinguish between a tone and no tone, as needed to interpret Morse, than to differentiate between different audio levels and tones, which is required to understand speech. More technically, a CW signal has a lower information content than speech, after all it's slower and there's no concept of accent or intonation that applies to speech. Accordingly, it occupies a smaller bandwidth, and this increases its SNR (signal to noise ratio). It's been suggested that CW has a 13-16dB gain in SNR over SSB, and that's equivalent to increasing the transmitted power by a factor of up to 40.

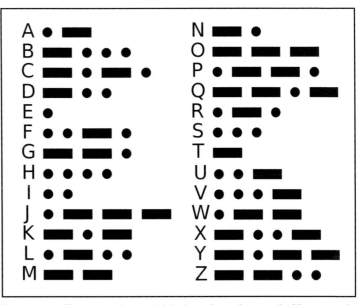

Figure 9.1. The use of a variable length codes made Morse code remarkably sophisticated, for something developed almost 200 years ago.

An Overview

As you're probably aware, Morse code allows letters, figures and punctuations to be transmitted by sequences of short and long sounds. Colloquially referred to as dots and dashes, respectively, because of the way they're written, radio amateurs tend to say 'dit' and 'dah' because these sound more like short and long sounds when spoken. We present the codes for the letters as **Figure 9.1** but, if you're not familiar with Morse, the apparently random nature of the code probably isn't what you'd have expected. Since character coding schemes like ASCII follow a pattern – so A is hexadecimal 41 (01000001 binary), B is 42 (01000010), C is 43 (01000011) etc. – it's reasonable to question why Morse doesn't follow a similarly logical sequence. The answer is that the codes were chosen so that the letters most commonly encountered in the English language are represented by the shortest codes. So, for example, E and T, the two most common letters, have the codes '•' and '-' respectively, while the lesser used letters have longer codes (e.g. '•---' for J and '--•-' for Q). This reduces the overall number of dits and dahs in a typical message, compared to the use of a fixed length code. This remarkable fact is commonly not appreciated, and the technique was largely forgotten until the Hoffman Code for data compression was introduced in 1952 and used much the same technique of variable length codes.

Sending and Receiving Morse

In order to transmit Morse you need to learn it, and at a reasonable speed – sending it by looking up each character before you send it just isn't an option, and receiving it using a look-up table would be nearly impossible. There's lots of information online, and plenty of resources to help you, but all recognised methods involve learning to receive Morse before you try to send it. After all, only when you've learned to receive it will you be able to send Morse with the correct rhythm, by which we mean the correct length of a dah with respect to a dit, and the correct lengths of the spaces between the dits and dahs in a character, and between the characters.

When you do start to send Morse, though, you need a Morse key, and there are several options. It's good practice to start with a manual key, otherwise known as a straight key, which is effectively a finely-balanced spring-loaded switch. Press the lever down for a short time to transmit a dit, and press it down for longer to send a dah. You can see a straight key in **Photo 9.1**, which also

Photo 9.1. A straight key is the simplest way of sending Morse and the method you should start with.

1: Operating

shows the simple arrangement to send an audio tone that you'd use while practising.

There are more sophisticated ways of sending Morse and you'll probably move onto one of these eventually if you get serious about Morse. First up is something called an automatic keyer, and it looks a bit like a manual key on its side. So, rather than pressing the lever down, you press it sideways. You don't have to judge the lengths of dits and dahs yourself, though. Press it left and it generates a string of dahs for as long as you hold it in that position; press it right and it generates dits. An automatic keyer comprises the key itself, and the electronics or software to generate the strings of dits or dahs. Going one stage further we come to dual-paddle automatic keyers, otherwise known as squeeze-keyers. Sending strings of dits and dahs is achieved in the same way as with a single-paddle keyer, but if you squeeze the two together you get a string of alternate dits and dahs. The number of movements decreases from manual to single-paddle keyers, and from single-paddle to squeeze-keyers. Accordingly, the maximum achievable speed increases in the same order. Do be aware, though, that although automatic keyers can reduce fatigue, you shouldn't be tempted to send Morse faster than you can receive it. After all, if you do, someone might well reply to you at the same speed. The final step in the evolution of methods of sending Morse is via software that allows Morse to be sent just by typing the required character. This really hasn't caught on to the same extent, though, probably because it takes away much of the skill that amateurs like to perfect.

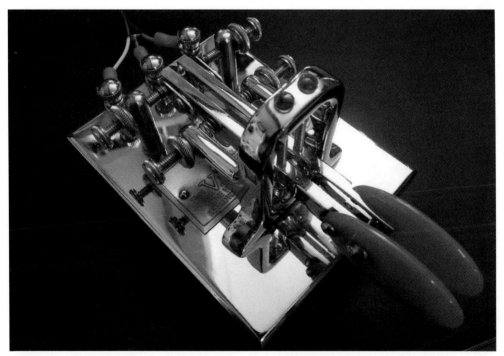

Photo 9.2. A double paddle key like this one is connected either to discrete electronics or something like a Raspberry Pi to generates strings of dits, dahs and alternating Dit-Dahs. Photo: BernieBLK.

Thing 10 – Employ State-of-the-Art Digital Modes

Having looked at a very early code for transmitting textual messages, namely Morse code, we now turn our attention to something a bit more modern. And here we don't just find a single code for transmitting and receiving text or other forms of data via radio, but several. Most of these codes are very much of the digital age, but in many cases they were being used by radio amateurs long before text messaging, WhatsApp and the like made their appearance.

In the Beginning

Although Morse represented the first mainstream method for transmitting textual information – first by telegraph lines and then by radio – it differs in one important respect from today's perception of text exchanges. That difference is that the text wasn't translated to code and vice versa automatically, but by a telegraph operator. Automated transmission and reception of text came in the form of an electro-mechanical device called a teleprinter, which looked rather like an overgrown typewriter. It also differed from Morse code in that, while Morse could barely be considered a binary code, teleprinters genuinely did use a binary code.

The very first teleprinters used a 5-bit code, undoubtedly because of the difficulty in using a longer code in an electro-mechanical device. However, that allowed only 32 different characters to be transmitted, and this is less than the total number of figures and letters, even if only capital letters were used. The solution was to use two character sets, one for letters and the other for figures and punctuation, and to use two control codes to switch between them. Mechanical considerations also limited the speed – 60 words per minute was typical. Radio amateurs still use the teleprinter's 5-bit code, and it's a rather pedestrian speed, in a mode referred to as RTTY (radio teletype), but why when the world has moved on? Part of the rationale is surely nostalgia, preserving our history. However, another factor is undoubtedly in a quest for variety. After all, why have one mode for data transmission when you can have several? What's more, each has its own features and benefits or, in some cases, quirks and idiosyncrasies. And to make that point, radio amateurs have also used a more modern mode that started life in the types of teleprinters, aka teletypes, that were some of the first devices to be used as a computer terminal. Using the 7-bit ASCII code – a derivative of which

Photo 10.1. RTTY, amateur radio's first truly digital mode, was based on the electromechanical teleprinters of old. Photo: Flygvapenmuseum.

1: Operating

LETTERS	A	B	C	D	E	F	G	H	I	J	K	L	M	N	O	P	Q	R	S	T	U	V	W	X	Y	Z	CARRIAGE RETURN	LINE FEED	LETTERS	FIGURES	SPACE	ALL-SPACE NOT IN USE
FIGURES	-	?	:	WHO ARE YOU	3	%	@	£	8 BELL	(	)	.	,	9	0	1	4	'	5	7	=	2	/	6	+						
CODE ELEMENTS 1	●	●		●	●	●			●	●					●		●		●		●	●	●	●			●	●				
2	●		●			●		●	●	●	●					●	●	●			●	●	●			●		●	●			
3		●			●		●	●				●	●		●	●				●			●	●					●		●	
4		●	●	●			●	●				●	●	●		●			●			●			●				●	●		
5	●					●	●						●	●		●	●	●		●			●	●	●	●			●	●		

Figure 10.1. Using a 5-bit digital code, RTTY had a limited character set and had to use shift codes. Based on image by Ali Lokhandwala.

is still used for web traffic today – it didn't need to shift between character sets. Radio amateurs refer to it simply as ASCII, it allows both upper- and lower-case characters to be sent, and it's somewhat faster than RTTY. However, it's largely died out today – unlike its less sophisticated predecessor RTTY – because there are far better modes available.

Moving On
The next generation of data modes was designed specifically for radio communication, where noise and interference are more likely than on telegraph lines. These modes were also designed to be implemented by software, which allowed them to be more sophisticated. The first such mode, called AMTOR, used a 7-bit code. However, only 35 of those possible 128 binary combinations are used, all of which have four 1s and three 0s. These were picked so an error in any single bit won't turn it into another valid code and, therefore, many errors could readily be identified. There are two operating codes. Before you actually make contact with another station, each character is sent twice to provide a degree of additional error protection. Needless to say, that halved the speed. Once contact is made, though, a different mode is used. Characters are sent three at a time, after which the transmitting station waits for the other station to confirm that it's received them without error. If such an acknowledgement is received, another three characters are sent; if it isn't, the three characters are repeated. AMTOR is rarely used today, partly because of the difficulties sometimes encountered in maintaining synchronisation between the two stations. But it does illustrate some of the features that are beneficial on less-than-perfect communication channels.

Perhaps the most popular mode today for conversational use is called PSK31. By conversational use we are talking of contacts in which an operator can type characters which are received in real time by another station. All the modes we're looking at here are conversational, but as we'll see in Thing 11, there is an intriguing alternative.

PSK31 was designed to transmit at a typical typing speed – after all, anything faster is pointless for conversational use – so it's actually a bit slower than RTTY when measured in words per minute. However, unlike RTTY or AMTOR, but in common with Morse code, PSK31 used a variable length code, with the shorter codes being assigned to the more-commonly encountered letters. The upshot off all this is that PSK31 only transmits at 31bps. Since PSK31 broke new ground in using phase shift keying, it was able to achieve a bandwidth of just 60Hz. This increases the number of stations that can operate in a given chunk of an amateur band, and it very much decreases the negative effect of interference. The upshot of all this is that PSK31 allows long-distance contacts to be made with low power.

Slow Scan Television

Figure 10.2. SSTV webcams show you SSTV images that have recently been received worldwide.

It was only our intention to present a few of the data modes available – after all there are just so many of them – so you can get a feel for them and the issues that have influenced their design. However, we really can't leave this topic before mentioning one more, because its purpose isn't to send textual information.

Here we're referring to SSTV (slow scan television), but it's not like regular television in that it's used for sending still images, not moving pictures. In that sense, it's not too dissimilar to the concept of sending a photo via email or WhatsApp. Certainly radio amateurs do transmit and receive moving TV signals (see Thing 43), but this is mostly in the 70cm (430MHz) band and above, because only at these frequencies is adequate bandwidth available.

SSTV isn't fast, and the resolution is rather limited, but do bear in mind that it's able to send a picture on the shortwave bands which, while being subject to noise, interference and fading, are able to span the globe. In the next section we'll provide some advice on how to re-

Photo 10.2a. SSTV permits images to be exchanged over the shortwave bands.

ceive SSTV signals, and other modes too. However, to see the SSTV pictures that radio amateurs around the world are receiving, take a look at www.worldsstv.com which shows recently received images and has links to several online SSTV cams.

Over to you

Hopefully, one day, you might find yourself transmitting textual and SSTV images around the world. For now, though, how about trying your hand at receiving these fascinating signals? After all, this could be the motivation you need to take the next step.

If you have a receiver that covers several amateur bands, all you need to start receiving data signals is some free software. MultiPSK (http://f6cte.free.fr) is a good choice for Windows, and Fldigi (www.w1hkj.com), while not supporting as many modes as MultiPSK, is also available for Linux. You will need to read up on which frequencies or parts of the various bands to monitor for the different modes, and listen to recordings so you can identify modes by ear. Ideally, you should wire the receiver's audio output to your PC's audio input. If your receiver doesn't have an audio output socket, you could use a microphone on your PC at a pinch.

If you don't have a suitable receiver, you could use a WebSDR, as discussed in Thing 1. However, this will normally involve you getting to grips with the software routing of the browser's audio output to the audio input of the decoding software. Alternatively, if you choose one of the WebSDRs that's based on KiwiSDR, you'll find that, while it doesn't support many other data modes, it has a built-in SSTV decoder. To find this, look in the Extensions menu.

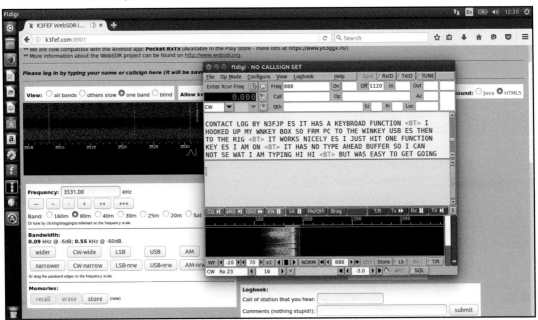

Figure 10.3. MultiPSK and Fldigi can be used to decode Morse code and many data modes. They can also be used with a WebSDR as shown here.

27

Thing 11 – Receive Signals you can't Even Hear

We saw in Thing 9 how Morse code allows a signal to be interpreted when a combination of a weak signal and interference would render speech almost intelligible. Here we go one better by looking at how you can receive signals you can't even hear or, conversely, transmit a signal that someone else can interpret without being able to it hear. As in Thing 10, we're looking at state-of-the-art data modes, but ones we've chosen to investigate separately because they are rather special. Welcome to the world of weak signal modes!

Non-conversational Modes

We used the term 'conversational modes' to describe the modes we look at in Thing 10. By this, we mean that they can be used to have an ordinary conversation with someone, but using text rather than speech. As such, they provide an experience similar to web chats. If the requirement for real-time communication is relaxed, though, communication can be achieved with lower power, in the presence of interference, and when hampered in all sorts of other ways. This is the purpose of the data modes we're looking at here and, in some cases, they allow signals to be interpreted even if they're so weak you can't even detect their presence by listening.

We're not going to give a blow-by-blow account of the characteristics of these non-conversational modes, or even list them all. We will discuss some of the basic principles, though, and describe just a couple so you can see what they're like to use. A common feature of these modes is that they're slow – very slow indeed – as we'll see later. This is a huge benefit for improving weak signal reception. As a result of this, though, the lengths of messages, and sometimes even what information you can provide, are limited. The other feature that makes it easier to allow weak signals to be decoded – because it allows better synchronisation between stations – is a restriction on when a station can start a transmission and when it must end. Fortunately, though, all of this is handled by the software.

JT65

JT65 was designed specifically for Moon-bounce communication (see Thing 48) but is now used more generally. It allows messages of up to 13 characters to be sent. Each such message takes 50 seconds to send, and messages start at the top of the minute, one station on even numbered and the other on odd numbered minutes. The clocks on PCs have to be accurately set up for this to work properly. Completing a minimal contact, with each station telling the other their callsign, location and a signal report, will take several minutes. But signals 24dB below the noise floor can be received. This figure is, perhaps, slightly misleading because it refers to a receiver's 2.5kHz bandwidth, rather than after it's been filtered in the software to JT65's 2.692Hz reception bandwidth. However, it does shed some light on why you're not going to hear weaker JT65 signals audibly. After all, that 24dB translates to the signal being 250 times weaker than the noise. And finally on JT65, the PSK Reporter website (https://pskreporter.info) allows users to see where their transmissions have been heard. This is compiled using real-time reports contributed

1: Operating

by many radio amateur, for several data modes.

WSPR

WSPR, which is pronounced 'whisper' for obvious reasons, also allows ultra-weak signals to be interpreted but is very different from JT65, for example. While most other modes are intended for two-way communication between two stations, WSPR has a very different role.

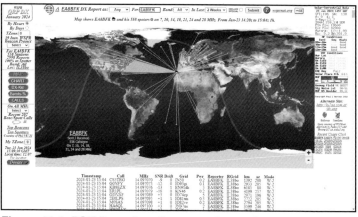

Figure 11.1. PSK Reporter allows you to monitor your transmissions in JT65 or other data modes too.

Those letters actually stand for Weak Signal Propagation Reporter, and it was designed to allow you to discover how far and where your signals can be heard. As it's alternatively been expressed, it allows you to test propagation paths. So why doesn't it permit two-way contacts? Basically, because it's been optimised to operate in a one-way beacon mode. In particular, it reduces the length of the message to include only what's necessary for beacon use. That information, which must be sent in a predefined sequence under the control of the transmission software, comprises just the callsign, the grid locator, and the power in dBm. When compressed, this data is just 50 bits long. It's transmitted over just less than two minutes with transmissions starting in even numbered minutes, and being continuously repeated. These characteristics result in improved performance over JT56; WSPR can operate down to 28dB below the 2.5kHz noise floor, and that's a 2.5 times improvement. Like other forms of data transmissions, you can monitor your signals on the PSK reporter (or anyone else's), but you might prefer to use the official WSPR site at https://www.wsprnet.org.

Figure 11.2. The WSPRnet website is the official repository for checking where your WSPR transmissions have been heard.

Photo 11.1. Raspberry Pi single board computers have been used at the heart of a WSPR beacon. Photo: Gerolf Ziegenhain.

29

Thing 12 – Learn about Propagation, Sunspots and all

Worldwide coverage is one of the unique aspects of amateur radio. After all, while your mobile phone might superficially appear to do that, in reality it only transmits to the closest base station. Having said that, you can certainly talk to a radio amateur in Australia without ever thinking of how your signal gets there. As someone with a technical frame of mind, though, you'll probably want to understand what goes on behind the scenes. And, although you could get that appreciation by reading a book, there's surely something rather special about discovering these things yourself. Even so, a bit of background information will help you to understand the basics of radio propagation, basics that you'll surely build on as you spend more time on the air. Here we're taking a look at the bread and butter of propagation in the shortwave bands. We're only scratching the surface though, so we're leaving you with plenty more to discover about shortwave propagation. We also look at two of the more esoteric propagation methods in Things 47 and 48, namely by bouncing signals off meteor trails and even from the surface of the Moon.

Line-of-Sight and Surface Waves

Later, we'll be looking at how some radio waves can reach the most distant parts of the world by bouncing between the earth and the ionosphere. This certainly isn't a universal model of propagation, though, so we need to start with something more fundamental.

Radio waves are usually considered to travel in straight lines and that's often a valid assumption. This can happen with radio in any part of the spectrum, and is the only mainstream mode of propagation for microwaves. However, if you're looking out over the sea, the horizon will be only about 5km away, perhaps increasing to 10km if you're on top of a 5m tower. That would seem to be a show-stopper for long distance microwave links, and this begs the question of how contact has been made over 114km on 241GHz (Thing 6). In part, such long distances are commonly spanned by extending the line-of-sight distance by operating from mountain tops. However, non-line-of-sight communication can be achieved on some frequencies using tropospheric propagation, although we're not going to get embroiled in that here. What we will say, though, is that the troposphere is the lowest region of the atmosphere, and its properties are affected by weather. And while it can provide over-the-horizon communication, it can't offer anything like the global range of ionospheric propagation, which we look at a bit later.

The other type of propagation we need to mention before we delve into the ionosphere, is called surface wave propagation, sometimes referred to as 'ground wave propagation'. It involves an interaction between the bottom of the transmitted signal's wave front and the Earth's surface. Because currents are induced in the surface of the Earth, the lower part of the wave front is slowed down, thereby causing the signal to tilt downward and thereby follow the curvature of the Earth – see **Figure 12.1**. The bottom line, therefore, is that communication beyond line-of-sight is possible. The range of the surface wave increases with decreasing frequency, so it can provide long-distance communication at VLF and LF. However, in passing, we should say that this is somewhat limited because of the difficulty of installing efficient antennas at these frequencies. Surface wave propagation is also feasible in the amateur MF and HF bands, albeit decreasing in efficiency as the frequency rises, and is largely ineffective at VHF and above.

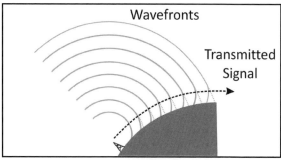

Figure 12.1. Interaction between the signal and the Earth's surface allows it to follow the curvature of the Earth. This makes communication possible beyond line of sight, especially at VLF and LF.

A Journey into the Ionosphere

So we've seen that surface wave propagation can allow over-the-horizon communication in the shortwave bands, but at these frequencies it's not going to provide global coverage. Indeed, range is typically limited to around 150km in the 160m (1.8MHz) band, reducing to little more than 15km in the 10m (28MHz) band. To see how global coverage is possible we need to turn our attention to sky wave propagation, which is responsible for long-distance communication on the shortwave bands. In order to understand this, though, we need to delve into the ionosphere.

The ionosphere is the region of the atmosphere that ranges from approximately 60km to 275km in altitude. It comprises gaseous atoms and molecules that are capable of being ionised by ultraviolet and other radiation from the Sun. And, of significance to our discussion, these ionised particles are able to interact with radio waves, as we'll see later. Getting into a bit more detail about the ionosphere, it's split into several layers which, in order of increasing altitude, are the D, E and F layers, of which the D layer is dissipated at night, and the F layer splits into the F1 and F2 sub-layers during the day. This is summed up in **Figure 12.2**.

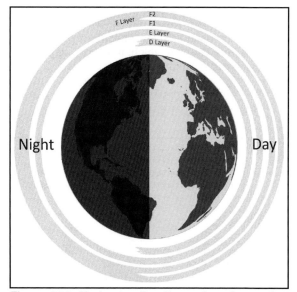

Figure 12.2. Understanding the various layers that make up the ionosphere is key to understanding propagation in the shortwave bands.

The other bit of background that we need to know about concerns a phenomenon that takes place about 147 million kilometres away on the surface of the Sun. We're talking here of sunspots, and these are regions of the Sun, each about the size of the Earth, which are cooler than the surrounding regions due to their much higher magnetic field. Sunspots increase the Sun's emission of high energy

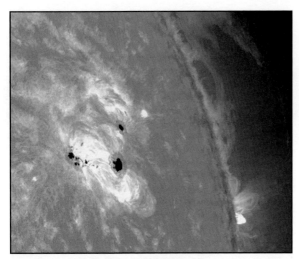

Photo 12.1. Ionisation of the ionosphere depends on radiation from the Sun which, in turn, is dependent on solar features called sunspots.

radiation that is responsible for ionisation in the ionosphere. The final bit of background information concerns the number of sunspots, and the bottom line is that the number varies in an 11-year cycle. At the troughs of these cycles there are very few sunspots, if any. But while we can say that there are more cycles at the peak of a cycle, not all cycles are equal in this respect. Suffice, to say, though, that the ionisation of the ionosphere will increase and decrease over an 11-year cycle, and this has a bearing on shortwave propagation.

Ionospheric Propagation

To see how the ionosphere has a major impact on propagation, let's look at each of the layers in the order a radio signal encounters them – see Figure 12.2.

The first such layer is the D layer, and its effect on radio waves is negative – it

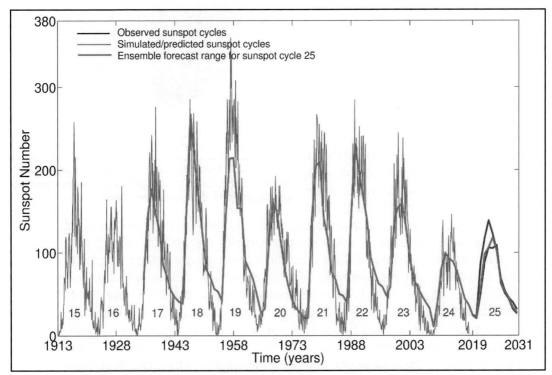

Figure 12.3. The number of sunspots follows an 11-year cycle, but not all peaks are the same height. Image: Idnan007. Image: NASA/SDO/AIA/HMI/Goddard Space Flight Center.

absorbs them. This effect is greater at the lower short-wave frequencies and can prevent radio waves from reaching the higher layers, which facilitate long-distance communication. Fortunately, though, the D layer is only weakly ionised so it disappears at night when the source of that ionisation, i.e. the Sun, can't affect it. Taken together, these facts explain why long-distance communication is not possible during the day on the 160m (1.8MHz) and 80m (3.5MHz) bands, but the conditions are more favourable at night.

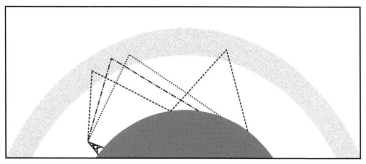

Figure 12.4. By refracting off the F Layer and, perhaps, reflecting back into the ionosphere, signals can reach the furthest parts of the globe.

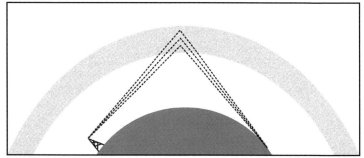

Figure 12.5. Long-distance shortwave communication is achievable by refraction from the F layer, but multi-path effects can cause distortion and fading.

Next we come to the E layer, and the bottom line is that it doesn't have a major effect on propagation in most of the shortwave bands, neither positive nor negative. However, it is responsible for so-called Sporadic-E propagation. This affects the upper-end of the HF band and VHF frequencies, and is one of the few mechanisms for propagation at much beyond line-of-sight at VHF. Indeed, ranges of up to 2,500km can be achieved on the 2m (144MHz) bands under such conditions. Unfortunately, though, as the name suggests, this form of propagation isn't common, nor is it easy to predict.

And so we come to the F layer, which is where most of the magic happens. Putting it simply, except in the case of near vertical angles, radio waves reaching the F layer can be refracted so that they return to the Earth's surface, some considerable distance from where they originated. Since signals will be launched at a variety of angles, this can give rise to signals being returned to Earth at a whole range of locations. And it gets better. That signal can be reflected from the Earth's surface back into the ionosphere, and this gives rise to even longer distance propagation via two or more hops – single- and dual-hop propagation is illustrated in **Figure 12.4**. It's possible, in this way, for radio signals to reach the opposite side of the globe.

However, it's not all plain sailing with long-distance HF propagation, and this gives rise to common perception – but by no means universally true – of short-

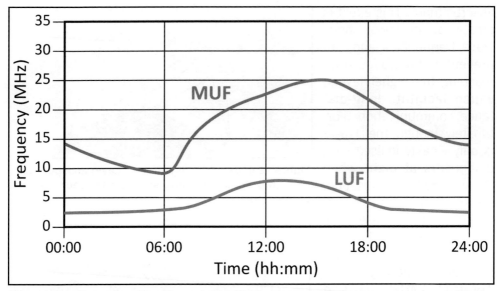

Figure 12.6. For any path, a lowest and Maximum useable frequency apply, and these depend on the time of day and the sunspot count.

wave radio. Any radio signal reaching the ionosphere will comprise rays over a range of angles which can, potentially, arrive at the same point via different paths – see **Figure 12.5**. The received signal can, therefore, be the sum of several signals, perhaps received at slightly different times, thereby distorting the signal. What's more, because the characteristics of the F layer can fluctuate over short periods of time, rapid changes in the phase difference between signals arriving along different paths can occur. This can result in constructive and destructive interference which will vary with time, this being perceived as periodic fading.

Needless to say, things aren't quite as simple as the last couple of paragraphs might suggest, and this brings us to the topic of the lowest useable frequency (LUF) and the maximum useable frequency (MUF). First of all, the LUF, and we've pretty much explained this already in our discussion of the D layer and the absorption it causes. Because that absorption decreases with increasing frequency, signals can only reach the F layer above a certain frequency. This is the LUF and, because the D layer dissipates at night, the LUF depends on the time of day. Turning to the MUF, we have to consider the fact that greater levels of ionisation are required in the F layer to refract a signal as the frequency increases. So, there's always an MUF, and this depends on the amount of ionisation. Because the degree of ionisation is greater during the day, the MUF is higher during daylight hours, but there's more. The level of F layer ionisation is very much dependent on the number of sunspots, and hence on the position within the sun's 11-year cycle. In practice, this means that the higher HF bands such as 15m (21MHz) and 10m (28MHz) especially, are largely unusable, perhaps for several years, around the troughs of a sunspot cycle.

1: Operating

Thing 13 – Take Part in a Contest

Amateur radio can be a challenging pursuit, as you'll have already seen as we looked at some of the grand challenges like transmitting at VLF or in the microwave bands. However, these challenges involve the radio amateur pitting his wits against nature. Here, though, we're going to see one aspect of competitive amateur radio and, in particular, taking part in a contest.

What is a Contest?

First of all we need to look at just what we mean when we talk about an amateur radio contest. The rules vary from one contest to another, but generally they're competitions in which each contestant makes as many contacts as they can during a fixed time period. Contests differ in which bands can be used, the modes that can be used (e.g. CW only, SSB only, data modes or mixed), the power levels allowed, whether contestants are single operators of whether teams are allowed, and how the contacts are scored. Some contests are universal; others are restricted to amateurs from specified geographical areas. There are also contests that are only for portable radio stations – usually called Field Days, but we look at these separately in Thing 30.

Many contests are arranged by national radio societies, often for amateurs in those countries, although some of the larger contests are unrestricted in this sense. If you become a contest enthusiast, you'll undoubtedly want to investigate what the RSGB offers. To cut a long story short, though, a large number of events are arranged through the year, many of which are open to amateur radio clubs, while several are of short duration, so they can provide a competitively easy route in contesting.

To find the world's largest contents, though, we have to cross the Atlantic where we find the CQ WW contests. Organised by CQ magazine, there are three contests, one for CW, one for SSB, and one other for RTTY contacts. Contestants can either be single operators or teams of multiple operators. And while multi-operator entrants can always operate on multiple bands, single operators have the choice of operating only in a single band. The other thing to bear in mind that CQ WW contests take place over an entire weekend, i.e. they're 48 hour events.

Photo 13.1. Many contests are open to single operators who will operate their normal stations. Photo: Giorgio Minguzzi.

35

50 Things for Radio Amateurs

Taking part in a Contest

First of all we should point out that taking part in a contest doesn't necessarily mean competing in a contest. Even if you're not competing, a contact with you will be just as beneficial to contestants as one with another contestant. Furthermore, since big contests very much increase the activity on the bands, they could well help you to make contacts with countries of other geographical entities you haven't hooked up with before. Do be sure to adhere to the contest rules of what information should be exchanged, though, and bear in mind that you shouldn't try to engage contestants in general chat.

If you intend on taking part, though, there's much more you should bear in mind. Much of this is obvious but here's a quick summary.

1. Make sure you check that your equipment is working properly beforehand, and that your antennas don't look like they're about to collapse.
2. Carefully read the contest rules beforehand and, in particular, be sure you understand what information you need to exchange for a contact to count.
3. During the contest, exchange the required information for each contact, speaking clearly, and logging the information according to the rules.
4. For long contests, make sure you have an ample supply of strong black coffee.
5. And after the contest, be sure you submit your entry within the required timescale.

World Radiosport Team Championship

It's certainly not for everyone – in fact you have to be invited to take part – but we thought it would be interesting to make brief reference to the pinnacle of contesting. The World Radiosport Team Championship (WRTC), which takes place every four years, is an on-site team competition that travels around the world. So that WRTC events are based on operating excellence and not external aspects, all teams are located in areas with the same propagation terrain and equipped with identical antennas. The next event will take place in the UK in 2026, utilising 50 sites in East Anglia, one for each team. Selection of participants will be made of the basis of a score that's calculated from their performance in specified contests in the run-up to the event.

Photo 13.3. The World Radiosport Team Championship provides the pinnacle of contesting achievement. Photo: R. A. Wilson.

1: Operating

Thing 14 – Link up with every Country in the World

Even if we ignore formal competitive events such as contests (Thing 13), many radio amateur do like to set themselves goals such as making contacts with as many of the world's countries as possible. If you get serious about this quest, though, there's certainly a more formally competitive aspect such as a position near the top of the so-called DXCC honour roll. Even if you don't aspire to such dizzy heights, though, there can be quite a buzz from making contact with that country you've been chasing for months or years, especially if that contact was made on a challenging band.

Introducing DXCC

DXCC stands for DX Century Club, with DX being amateur radio speak for long distance. It's actually an award (see Thing 16) which amateurs can apply for if they qualify. The basic award is for making contact with 100 countries, on single bands or with single modes, or on any or all bands using any modes. Stickers to endorse the basic award are issued as you make contact with more countries.

The ARRL, the American Radio Relay League, America's national amateur radio society which administers DXCC, also maintains an honour roll which is hosted on its website. And as this is being written, it shows that top place being held by Israeli radio amateur Ami Shami, 4X4DK. But here's an intriguing fact – the ARRL lists 340 countries. And as you might have worked out, that's more than the generally accepted number of countries in the world. It looks as if we need to investigate this in a bit more detail.

What's a Country?

According to the Nations Online website, there are 195 independent sovereign nations in the world, not including the disputed but de facto independent Taiwan, plus some 60 dependent areas and several disputed territories, such as Kosovo. Already we can see how this question can have no clear answer. But even if we add the 60 dependent areas and disputed territories to that figure of 195, we don't get close to 340 which is the maximum number of available countries according to the ARRL.

To be fair, although many

Figure 14.1. Bizarrely, radio amateurs can contact many more countries than those generally recognised as such.

radio amateurs consider DXCC to be an award based on contacting countries, the ARRL actually calls them by the even more confusing word 'entity'. Admittedly, though, they do define the criteria by which an entity gains a position on their list. It probably wouldn't be too exciting to look at all the criteria by which entities qualify for the list, but to give you an idea, let's consider a few examples. The United Kingdom is a sovereign state and a member of the United Nations, but it doesn't appear as a DXCC entity. Instead, England, Scotland, Northern Ireland and Wales are listed separately. So too are the Isle of Man, Jersey and Guernsey, even though these have a rather different status. And to give another example, islands are often separate entities, but only if they are 350km from their parent country. So, for example, we could mention the Canary Islands which are part of Spain, and Madeira which is part of Portugal. Perhaps it all seems a bit arbitrary, but if you're looking for a challenge, those 340 entities certainly provide it.

As a parting shot, as an alternative to making contacts with DXCC entities, how about taking a trip to some of the rarer ones with your radio gear on a DXpedition (see Thing 31)? You could find yourself very popular. And if you want some inspiration, North Korea, Scarborough Reef and San Felix Islands are reportedly the three most sought after. We do have to say, though, smuggling amateur radio gear into North Korea certainly isn't recommended.

Photo 14.1. Chances are you've never heard of the French sub-Antarctic Kerguelen Islands. You'd be much in demand if you set up a radio station there though.

Thing 15 – Use DX Clusters and Propagation Websites

At one time, discovering that a station in a much sought-after location was on the air was mostly a matter of luck, unless you spent your entire life on the bands. I guess there was a possibility that a friend might give you a heads-up by phone, but the chances are they'd be busy trying to make contact themselves. All that has changed, thanks to DX Clusters and propagation websites.

DX Clusters

A DX cluster is a website that provides real-time information about which radio amateurs can be heard on various bands. This is usually presented as a table, with the most recent entry at the top, and with columns for the time, callsign of the spotter (i.e. the station making the report), the callsign of the station spotted, the frequency, and perhaps a comment. Some provide separate tables for each band. Others show all bands, but commonly allow you to apply filters for a particular band, and perhaps also for the mode of interest, i.e. CW or SSB. Occasionally the country of the spotted station is shown, or sometimes this is indicated by a flag. Otherwise, you'll need to know callsign prefixes to identify the location of stations, or be able to look them up quickly. Some DX clusters allow you to carry out searches, most commonly for a particular callsign, but rarely is it possible to view only those with a spotter close to your location. Of course, that is what you'd really be interested in, so you'd often have to search by eye. Don't assume, though, that spotters will be local to the organisation hosting the DX cluster as that is rarely the case. A better view of what has been spotted from where can be obtained by consulting one of those DX clusters that offer a map view of recent spots. Rather than recommending any particular DX cluster, we suggest you search online with the view

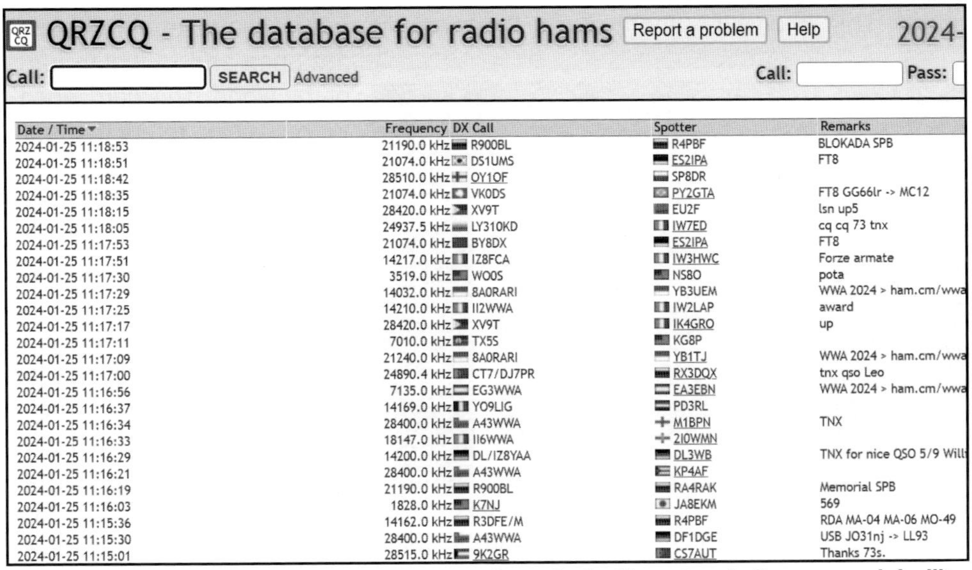
QRZCQ - The database for radio hams | Report a problem | Help | 2024-

Call: SEARCH Advanced   Call: Pass:

Date / Time	Frequency	DX Call	Spotter	Remarks
2024-01-25 11:18:53	21190.0 kHz	R900BL	R4PBF	BLOKADA SPB
2024-01-25 11:18:51	21074.0 kHz	DS1UMS	ES2IPA	FT8
2024-01-25 11:18:42	28510.0 kHz	OY1OF	SP8DR	
2024-01-25 11:18:35	21074.0 kHz	VK0DS	PY2GTA	FT8 GG66lr -> MC12
2024-01-25 11:18:15	28420.0 kHz	XV9T	EU2F	lsn up5
2024-01-25 11:18:05	24937.5 kHz	LY310KD	IW7ED	cq cq 73 tnx
2024-01-25 11:17:53	21074.0 kHz	BY8DX	ES2IPA	FT8
2024-01-25 11:17:51	14217.0 kHz	IZ8FCA	IW3HWC	Forze armate
2024-01-25 11:17:30	3519.0 kHz	WO0S	NS8O	pota
2024-01-25 11:17:29	14032.0 kHz	8A0RARI	YB3UEM	WWA 2024 > ham.cm/wwa
2024-01-25 11:17:25	14210.0 kHz	II2WWA	IW2LAP	award
2024-01-25 11:17:17	28420.0 kHz	XV9T	IK4GRO	up
2024-01-25 11:17:11	7010.0 kHz	TX5S	KG8P	
2024-01-25 11:17:09	21240.0 kHz	8A0RARI	YB1TJ	WWA 2024 > ham.cm/wwa
2024-01-25 11:17:00	24890.4 kHz	CT7/DJ7PR	RX3DQX	tnx qso Leo
2024-01-25 11:16:56	7135.0 kHz	EG3WWA	EA3EBN	WWA 2024 > ham.cm/wwa
2024-01-25 11:16:37	14169.0 kHz	YO9LIG	PD3RL	
2024-01-25 11:16:34	28400.0 kHz	A43WWA	M1BPN	TNX
2024-01-25 11:16:33	18147.0 kHz	II6WWA	2I0WMN	
2024-01-25 11:16:29	14200.0 kHz	DL/IZ8YAA	DL3WB	TNX for nice QSO 5/9 Will
2024-01-25 11:16:21	28400.0 kHz	A43WWA	KP4AF	
2024-01-25 11:16:19	21190.0 kHz	R900BL	RA4RAK	Memorial SPB
2024-01-25 11:16:03	1828.0 kHz	K7NJ	JA8EKM	569
2024-01-25 11:15:36	14162.0 kHz	R3DFE/M	R4PBF	RDA MA-04 MA-06 MO-49
2024-01-25 11:15:30	28400.0 kHz	A43WWA	DF1DGE	USB JO31nj -> LL93
2024-01-25 11:15:01	28515.0 kHz	9K2GR	CS7AUT	Thanks 73s.

Figure 15.1. This DX cluster, at www.qrzcq.com, has a typical layout and offers a search facility.

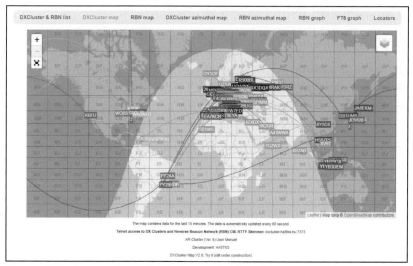

Figure 15.2. As an alternative to the usual list view, a few DX clusters like this one by HA8TKS display the information on a map.

of finding one that appeals to you. We should point out one potential downside of DX Clusters. Bear in mind that you won't be the only amateur consulting your chosen DX cluster. So, even if it only takes a few seconds to tune to the reported frequency, by the time you're on frequency you could well find that dozens of others are also trying to contact that station.

Real-time Propagation Websites

While DX clusters provide very specific information on what stations have been heard from where, an alternative online resource might be of interest. We're talking of websites that provide real-time quantitative information about several factors that influence propagation. You'll have to read up on just what they mean and how to interpret them, but factors aren't limited to just a sunspot count, and include other measures such as the A Index and the K Index. Some also provide maps which show, for example, the MUF (see Thing 12) in various parts of the world. This is possibly more useful than raw figures, in showing which bands are open and to which parts of the world. The word 'open', in this sense, refers to propagation conditions favouring communication to particular localities. Of course, just because a band is open doesn't mean that you'd find especially interesting stations, which is what you can get with a DX Cluster. So, it might be appropriate to consider DX clusters and real-time propagation as complementary resources that you could use hand-in-hand.

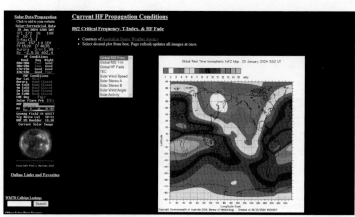

Figure 15.3. Real-time propagation sites – such as this one at www.hamqsl.com – can provide more generic information than DX clusters.

1: Operating

Thing 16 – Obtain Awards

We made passing reference to awards in Thing 14 as part of our discussion on contacting every country in the world. However, we thought the subject warranted more comment since it's an important part of amateur radio for many amateurs.

Awards are usually certificates, although they could be trophies, that recognise an achievement in amateur radio operation. So, for example, we've already seen the DXCC award that's given in recognition of making contact with at least 100 ARRL entities, commonly thought of as countries. Many awards, but by no means all, are given on the basis of making contact with all, or a specified minimum number, of political or geographic areas. And there's clearly some correlation between obtaining some of the more prestigious awards and an amateur's skill as a radio operator. Some of these awards are quite aesthetically pleasing, and are often framed and displayed with pride in the operator's shack. The word 'shack', if you're not familiar with it, isn't necessarily a wooden shed in the garden, but is amateur radio-esque for the building or room that houses the station. We can't say a great deal more generically, because details differ so much from one award to another. However, if this has sparked some interest, here are a few examples to fuel your curiosity.

Notable Awards

Continuing on the global theme of DXCC, other prestigious awards include the Worked all Continents (WAC) and CQ Worked all Zones (WAZ) awards. Having used another bit of amateur radio jargon, we should explain that working someone means making a two-way contact with that station. The first of these is self-explanatory, and perhaps doesn't seem too demanding, but that would be forgetting the possible endorsements. For example, a WAC for contacts on the 160m (1.8MHz) band would be much more challenging, whereas the ultimate is surely the Five Band WAC which requires all continents to be worked on 80m (3.5MHz), 40m (7MHz), 20m (14MHz), 15m (21MHz) and 10m (28MHz). The WAZ award requires contacts with all of the 40 zones into which the world was divided by CQ magazine who administer the award. While not global, another of the much sought-after awards is the Worked all States, those states being the ones that constitute the United States of America.

Figure 16.1. WAC Award Certificate

Figure 16.2. CQ Award Certificate

Here in the UK, the RSGB has an extensive awards programme. Several of these awards are designed to encourage newcomers into amateur radio. Included here are youth awards, plus schemes that are limited to Foundation and Intermediate level licence holders. Following in the footsteps of the DXCC award, the RSGB has an award concerned with

Figure 16.3. Islands on the Air Award Certificate

countries, in particular those countries in the British Commonwealth. Another especially interesting range of awards that the RSGB supports are those that constitute the IOTA (Islands on the Air) programme. Moving beyond the RSGB, another notable British awards programme is Worked All Britain (WAB). The most well-known WAB award is issued to amateurs who have worked stations in prescribed numbers of the 10km x 10km squares that Britain is divided into on Ordnance Survey maps. Like DXCC, which encourages people to activate rare DXCC entities, WAB encourages operation from rare squares.

Many of the world's national amateur radio societies or magazines operate award schemes. In many cases these are issued for contacts made within those countries so, commonly, they are especially popular with amateurs in those countries. However, some – like the CQ USA-CA (USA Counties Award) which is granted for contacts in specified numbers of US counties – have a more global appeal. And since there are no fewer than 3,077 such counties, qualifying for the USA-3077 award could keep you busy for quite some time. Other international awards have proved popular because the associated certificates are considered particularly attractive. Classic examples are some of the European award certificates that incorporate coats of arms of the respective administrative regions.

Award Application

The exact process of applying for an award differs but usually requires documented evidence of making the necessary contacts. Commonly that documentation consists of QSL cards – see Thing 17 – and there are various ways these could be provided to the awarding body. This might involve sending them by post, even though that runs the risk of loss, or it might be possible to submit scans, or use the services of an approved card checker. The use of various electronic alternatives may be permitted or, in a few cases, self-certification may be allowed. One other generic thing you need to know is that a fee will invariably be payable, and this might not be an insignificant amount.

  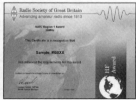 

Figure 16.2. The RSGB has a wide range of awards

1: Operating

Thing 17 – Collect Mementos from around the World

We could have entitled this "thing" as collect "QSL cards". But, as some of you aren't engrained in amateur radio speak, we thought we'd spell it out in plain English. As one of the Q-codes used in amateur radio, the official meaning of QSL is 'I confirm receipt of your transmission'. However, it's taken on a rather special meaning in amateur radio, especially when it's suffixed by the word 'card'.

Spotlight on QSL Cards

In the context of amateur radio, a QSL card is approximately the size of a postcard (remember those?), and is sent by one amateur to another to confirm that a contact took place. QSL cards were first sent in the dark ages, but they remain popular to the current day. Cards might be collected purely as mementos, but don't forget that they are also needed, in many cases, in applying for awards (Thing 16). A card could either be sent speculatively, in which case it would be filled in and sent shortly after the contact, or it could be sent in response to a card from the other party. In the latter case, the QSL card would only be issued after the amateur looked up and was able to confirm that the contact had genuinely occurred.

QSL cards can be of a standard design, and are available from companies that specialise in printing such cards. In this case, it'll be customised only by the addition of your callsign and whatever indication you choose to provide about your location. Alternatively, you can design your card from scratch and get them printed by a mainstream supplier and this isn't at all difficult if you use DTP or similar software. A totally customised design could show a photo or illustration of some sort, and this will commonly relate to your location. Cards can be single-sided, but it doesn't cost much more for them to be double-sided. On a double-sided card, the front will generally be an attractive design with your callsign and perhaps location, while the back will include more details, much of which will take the form of various boxes into which you'd fill in contact-specific details. These details will include the callsign of the station you contacted, time, date, frequency, mode and signal report, as a minimum. Before you decide on the design, though, be sure you take a good look at several sample cards to get a better idea of what they typically include.

Sending QSL cards via the postal service in the normal way can be very expensive. So there's an alternative means which is favoured. This is to employ the services of a QSL bureau. Typically operated by national radio societies as a

Photo 17.1. This QSL card might have been sent back in 1951 but exchanging such mementos from around the world remains popular. Photo: OE5HL.

service for their members, these bureaux make a huge cost saving by shipping cards in bulk. So, for example, you'd save cards up and send them to the bureau when you've reached a certain number, and you'd receive cards back in bulk in pre-paid envelopes that you provide. And between these extremes, cards are shipped from one bureau to another around the world, again in bulk. The downside of this is that it will take very much longer for you to receive QSL cards, perhaps months or even years.

The Electronic Alternative

The practice of sending QSL cards – or perhaps we should say physical QSL cards – is dying out. This is surely a response to higher postage costs and/or long waits between contact and QSL, and the fact there are now digital alternatives. So, what are those alternatives?

First up is LoTW (Logbook of the World) which is operated by the American national amateur radio society ARRL. It's a free service but requires registration. Users upload details of their contacts to LoTW. Whenever a matching report is detected – i.e. details are found from both participating parties in a contact – an electronic QSL is issued to both amateurs. These electronic QSLs can be used in applying for awards, but that's dependent on the policies of the various award bodies. Currently, LoTW is the only form of electronic QSL recognised by ARRL. Since ARRL is the body that issues some of the most prestigious wards such as DXCC, WAC and WAZ, this draws a lot of users to LoTW.

Not an imaginative name, but entirely appropriate, our next QSL card alternative is eQSL (www.eqsl.cc). The service is only available to registered users, although there is no compulsory fee. The major difference between eQSL and LoTW is that, while the latter is just a database, the former also allows users to design and send a graphical QSL 'card', albeit electronically. It rather seems that most users design their card on one of eQSL's standard templates, so there's a high degree of sameness. However, a fully customised design can be used, but only if you make a voluntary financial contribution. If you're an avid QSL card collector, the availability of a graphical image could well be of interest. After all, these cards can be printed out on good quality card and either put in an album or stuck on the wall, just like QSL cards of old.

And so we come to QRZ Logbook, which concludes our look at what are probably the big three. This is another free service for registered users, although additional features are available for a fee. You shouldn't read anything negative into this brief mention, but we will say that it's similar to LoTW and we don't want to repeat ourselves.

As a final comment on electronic QSLs, we suggest you read up on each of the solutions in some detail before coming to a decision. Things to consider include which are accepted by the various awards bodies, and what options are available for exchanging log data between platforms. Bear in mind, though, that all are free, so there's no financial loss in using more than one, although there will be a time penalty.

Thing 18 – Operate a Station Remotely

In Thing 1 we looked at WebSDRs or, in other words, radio receivers that are hosted on the internet. These allow anyone to listen on the amateur bands, and more, without having to buy a high-performance receiver. Here we go one further, by investigating amateur radio stations that are hosted on the web. But while WebSDRs only allow remote users to listen, remotely-operated amateur radio stations offer transmit access too, albeit only for suitably-licenced amateurs. There are two aspects to this. First, it's possible to operate stations that were set up by a third party, and second you can remotely operate your own station. Here we present just enough information on both these topics to show what's possible and to whet your appetite. Note also that we're only discussing what's possible from a technical viewpoint. Licensing issues also need to be addressed, and these may differ depending on which country issued your licence, and which country the remote stations are located in.

Third-party Stations

Remotely-controllable stations are often set up by amateur radio clubs for exclusive use by their members. However, individuals also make their stations available online, sometimes to any licensed amateurs. Commonly, such access is provided via www.remotehams.com because this removes the responsibility of the remote station owners to ensure that users are licensed. This free subscription service acts as a portal to a couple of hundred stations around the world, these being a mix of open-access stations, and those that have restricted access. We can imagine that this service could be of interest to radio amateurs who live in houses with limited space for antennas, or in apartments where antennas are prohibited. Having said that, though, it doesn't allow amateurs to build their own stations and antennas, technical aspects that are so much part of the hobby.

The sentiment that the best things in life are free doesn't necessarily apply to third-party stations though. So if you want the best, you might need to head over to www.remotehamradio.com. We mention this to show what's possible, and what might appear in other countries, but it's currently available only to American licence holders. In a nutshell, though, it provides access, for an annual subscription fee plus an airtime fee, to what we might describe as 'mega-stations'.

Figure 18.1. Remotehams.com offers remote access to amateur radio stations worldwide.

Remotely Operate your own Station

Remotely operating a third-party station has obvious benefits, but why would you want to operate your own station remotely? Well, there are several benefits. For example, imagine you're away from home and discover on a DX cluster (Thing 15) that a station is on the air in a country you've been trying to work for years. If you can operate your station remotely, you have the same opportunity to make that DX contact that you'd have if you were at home. It doesn't have to be just for DX though. Many amateur radio clubs arrange a club chat – or a 'net' as it's commonly called – at a particular time each week. So, just because you're stuck in a hotel room a couple of hundred miles from home, doesn't mean you can't catch up with you mates at the club.

And there's more. You'll recall that we suggested that remote third-party stations are useful for those who can't install antennas at home, and the same is true of remotely operating your own stations. However, that obviously depends on you having access to a remote location where you can set up that station. So this isn't for everyone, but if you have a holiday home, or if you have a good friend who owns a farm in the middle of nowhere, this could be exactly what you've been looking for.

To operate a third-party remote station, you don't need anything more than a laptop with a microphone, or even a mobile phone. If you're going to be using your own station remotely, though, you'll be responsible for setting up the necessary infrastructure at your remote station. This need not be too difficult, though, as you can see, as a block diagram, in **Figure 18.2**. We should point out, though, that it won't always be this simple. We're assuming that your transceiver has a USB port that can carry both control signals and digitised audio. If you have an older rig with a legacy control port, though, you'll need an interface to convert between the laptop's USB and your rig's control port. You'll also need to route audio signals via the laptop's sound card. And, needless to say, if you also want to be able to switch between antennas and/or rotate antennas, again you'll need more hardware.

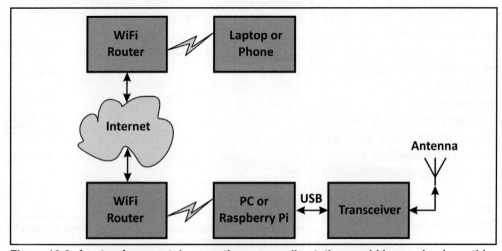

Figure 18.2. A setup for remotely operating your radio station could be as simple as this.

2

Construction

Introduction

Electronic construction, hardware development, making, hacking, call it what you will, but amateur radio can't be divorced from these important skills. What's more, electronics as a hobby has made a major comeback in recent years, thanks in no small part to the maker revolution and the birth of hackspaces where makers can meet like-minded people, learn and collaborate. So, if you consider yourself an electronics enthusiast, it's totally appropriate for us to show you how amateur radio could help you develop your hardware skills. So, here were going to delve into some of our 50 Things you can do with Amateur Radio which are concerned with electronic construction.

If most of your recent electronics construction has been concerned with interfacing to single-board computers like the Raspberry Pi, amateur radio can help you learn a whole lot more. So, we'll start by introducing you to

how radio communications requires hardware skills you might not be familiar with. We'll also explain how amateur radio is unique in allowing you to transmit using home-built equipment, something that's not permitted with licence-free devices like PMR walkie-talkies or CB radios. Then, we'll investigate a half-way solution to building your own gear in the form of assembling off-the-shelf kits, before conceding that the simplest amateur radio gear can be almost unbelievably simple to build from scratch. In these first few cases we'll include a practical element in suggesting some simple constructional projects, some of which you can use on the air. Next up, and bearing in mind that reliving the early days of electronics is popular among the maker community, we'll delve into using archaic, but still serviceable, hardware such as thermionic valves, aka electron tubes.

We then shift gear by looking at home-made antennas, a vitally-important part of any amateur radio station. These aren't built, primarily, using a soldering iron, a circuit board and a handful of components, so other skills come to the fore. Here we present several designs that you can experiment with. Also, taking you out of the house, we'll describe how you can set up a mobile amateur radio station, in other words one you can operate from your car.

Collaboration is a key element in amateur radio, so our next topic looks at how you can use your electronics expertise to engage in a much more ambitious project. In particular we'll outline the hardware of an amateur radio repeater. This leads into a discussion of how you could join a group with the aim of designing, building and operating a repeater station to improve the range of low-power or mobile stations.

And to round up our investigation of making your own gear we turn our attention to software, something which, increasingly, is closely tied up with electronics hardware. So, just as amateur radio provides the ideal opportunity to expand your hardware design and construction skills, we then see how it also provides you with the prospect of brushing up your coding skills. Specifically, we'll investigate how you could code a software-defined radio receiver. And, finally in this chapter, we'll turn our attention to the Raspberry Pi, the world's best-selling single-board computer. Justifiably popular in the amateur radio community, we'll see how radio communication opens up so many new and exciting RPi applications.

2: Construction

Thing 19 – Expand your Electronics Construction Skills

Perhaps your recent experience of electronics has been associated with attaching switches, LEDs, sounders and the like to Arduino or Raspberry Pi computers. If so, we have news for you. While computer interfacing will have provided you with a good introduction to several areas of electronics design and construction, these are just a start. In particular, you probably won't have had much exposure to analogue circuitry because it's not nearly as commonly encountered as digital circuitry for interfacing to single-board computers. In amateur radio, though, analogue electronics continues to play an important role, despite the undeniable advances made in the realm of digital techniques.

Photo 19.1. Although you can build an amateur radio station entirely from commercial gear, home constructions remains a popular option.

RF Circuitry

To give you a feel for the new electronics challenges presented by radio equipment, let's think about a very simple radio receiver, as shown as **Figure 19.1**. At this level it's easy to understand. The antenna captures radio signals and converts them to minute voltages. These tiny signals are first processed by the RF stage (RF stands for radio frequency) which selects the wanted signal. In other words it extracts one particular radio station, as selected by the tuning knob, from the myriad of other signals. In particular, it selects any signals on a chosen frequency. This is then passed on to something called the detector, which extracts the audio – typically speech – from the radio signal on which it was superimposed for transmission. This audio signal is then routed through to an AF amplifier (AF stands for audio frequency) which boosts it to such a level that it can drive a speaker. To get a feel for how different radio circuitry is from most digital electronics, we'll look in slightly more depth at a couple of the stages in our block diagram.

The first of these blocks is the one labelled "RF Stage", and its purpose is to

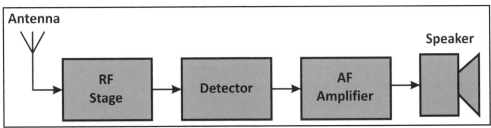

Figure 19.1. This circuit of a super-simple radio receiver allows us to get a glimpse of some of the components and circuit elements that will be largely unfamiliar if your experience is of the electronics for computer interfacing

Photo 19.2. Inductors, important elements in analogue radio circuitry, can either be air cored like this one, or wound on a ferrite core. Photo: CatalpaSpirit.

Photo 19.3. Variable capacitors are essential components in conventional analogue radio kit, in allowing the operating frequency to be selected. Photo: Elcap.

select signals on a single frequency by blocking all signals on other frequencies. This is accomplished using a so-called tuned circuit and, in particular, a parallel tuned circuit. This bit of circuit comprises a capacitor and an inductor, the values of which define the selected frequency. And if the selected frequency is tuneable, as it needs to be if you want to be able to choose which frequency and hence which radio station to listen to, the capacitor should be a variable capacitor. Let's look at these two potentially unfamiliar components. An inductor, otherwise known as a coil, is exactly what this latter name suggests, namely a coil of wire. Sometimes you'll be able to buy an appropriate coil off the shelf; in other cases you'll need to wind it yourself using enamelled copper wire. It can take the form of either a self supporting air-cored component, or one on a plastic tube, or alternatively one wound on a ferrite core. Turning to the variable capacitor, this is a fairly chunky mechanical device in which one set of metallic plates is moved with respect to another set of interleaved plates as the spindle is turned. That spindle is attached to a tuning knob that's usually mounted on the radio's front panel.

The other block that warrants some comment is the one labelled "Audio Amplifier". An amplifier, often comprising just a single transistor and a resistor, is commonly encountered in computer interfacing. This is needed if you're interfacing something to your computer which draws a higher current, or requires a higher voltage, than the computer can provide. However, that amplifier is digital, in the sense that its output will only ever be one of two voltages, for example 0V and +5V. This type of amplifier isn't appropriate for boosting the signal from the detector of our simple receiver, though, because speech is an analogue signal rather than a digital one. Accordingly, an analogue amplifier must be used. Depending on how much audio power you want, your amplifier might require only a single transistor, a resistor and a smattering of capacitors, but there's an easier way. That solution is to use an audio amplifier IC, these are commonly 8-pin, 14pin or 16-pin chips.

2: Construction

They're readily available and they're cheap.

Construction Techniques
Not only are the types of circuits used in radio equipment different from those encountered in digital applications, but the methods of prototyping and construction will often differ too. Breadboards have become very popular in recent years for prototyping circuits, perhaps before migrating the circuit to something more permanent such as a PCB. But, although their use streamlines the process for simple digital circuits, using a breadboard isn't always ideal for RF circuits. The first snag is that breadboards have long rows of parallel copper strips hidden away inside them, so they will exhibit capacitance, even when no components are plugged into them. If you now recall how tuned circuits, which comprise an inductor and a capacitor, are frequency specific, it starts to become clear how this stray capacitance could affect the working of the circuit. What's more, the problem is greater at higher frequencies. This is because the capacitors in tuned circuits will be smaller and, therefore, the additional capacitance that's internal to the breadboard will be more significant. The other issue is that ground connections are more important in RF circuits, and the thin copper strips in breadboards might not be adequate. The bottom line is that you might not find problems at lower frequencies – although it's surely preferable not to take the risk – but at VHF frequencies and above, you really ought to consider an alternative prototyping method.

While it's not nearly as easy as plugging component leads into breadboards, a prototyping technique commonly used for RF circuits involves the use of un-etched copper-clad laminated board, of the type which are used for producing printed circuit boards. The first thing to note about these boards is that they have a complete copper layer on at least one side, and this addresses the issue of poor ground connections with breadboards. The leads of any components that have a ground connection are soldered to the copper layer, and the other ends are soldered together in the air above the copper plane, keeping the leads as short as possible to reduce stray capacitance. The end result doesn't look pretty (see **Photo 19.4**), and there's the additional hassle of having to use a soldering iron, but good results are reported.

Not only do RF circuits require you to reappraise your prototyping methods but, if you want to build something more permanent, you might also need to rethink the de-

Photo 19.4. So-called 'dead bug' prototypes are rarely works of art, but they are much more likely to result in a working prototype than using a breadboard.

sign of the printed circuit board (PCB). Such a design might require a full ground plane – that is a full copper layer on one side of the board onto which ground connections are made – but, if you get up to UHF or microwave frequencies, there's a whole lot more you'll need to get to grips with. The intricacies are much too involved to get embroiled in here, but it does just hint at the voyage of discovery that you could set out on.

Getting Practical

So much for the basic principles, but how about getting practical? If you'd like to take your first steps in RF electronic construction, we have a couple of recommended projects for you.

The first project is really simple and quite an eye-opener. Called a crystal radio, this is the simplest possible radio receiver, it dates back to the birth of radio broadcasting in around 1920 and, bizarrely, it doesn't require a power supply. Instead the headphones are driven directly from the minute voltages that appear on the antenna. So, it's effectively the same as the simple radio receiver we looked at earlier except that is doesn't have the audio amplifier and, instead of the speakers, it drives a pair of high-impedance headphones or an earpiece. Here we introduce you to a particularly simple crystal radio – its circuit diagram appears as **Figure 19.2**. Stray capacitance isn't an issue, but using a breadboard isn't recommended because several of the components won't plug in easily. Instead, we suggest mounting the components on a wooden board so it looks the part – you could build it as shown in **Photo 19.5**. Probably the one thing we do need to clarify is that you should only connect one end of the coil to the variable capacitor. The connection between the other end of the coil and the capacitor should take the form of a flexible wire lead terminated by a crocodile clip so you can connect it to any tap.

The schematic doesn't show all component values, so here's what you need to know. The variable capacitor isn't critical, but somewhere between about 350 and 500pF will work well. Ideally, get a dual-gang variable capacitor, because it'll

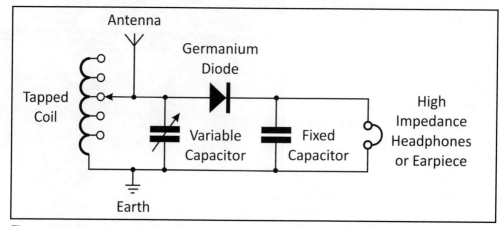

Figure 19.2. Here's a possible first RF construction project, and this crystal radio couldn't be simpler.

allow you to try further circuit configurations later. These will often be shown, for example, as 2 x 350pF which would mean two 360pF variable capacitors that are tuned together. You also need a plastic knob for the variable capacitor, because if you touch its spindle with your fingers you'll alter its tuning. The diode must be a germanium diode, instead of the much more common silicon diodes, possible part numbers being OA90, OA91 and 1N34A. The headphones or earpiece must be high impedance, so this excludes nearly all modern headphones. High impedance earpieces are available on eBay. The fixed capacitor should have a value of about 1-10nF, but this isn't critical.

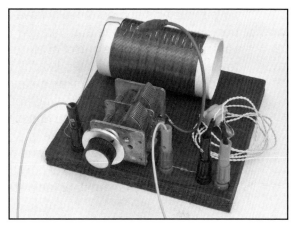

Photo 19.5. Building a crystal radio is simple enough but it introduces you to some key concepts in RF design.

The inductor is made by winding a length of enamelled copper wire around a plastic tube. Use a 55mm diameter tube and 22 SWG (Standard Wire Gauge) wire which is about 0.7mm in diameter. Use 100 turns, which is probably more than you actually need, but it's easier to use only part of the coil than to add more turns later. You should add connections, called taps, after every ten turns or so. To start, drill two small holes near one end of the tube and thread the end of the wire through them to secure it in place while you wind the wire. Keep the turns close together, and make a twist wherever you want a tap. If you want to take a break from winding, secure your work with insulating tape so it won't unravel before you return to it. When you've finished, secure the other end by threading it through another pair of holes. Finally, carefully use a very sharp knife (e.g. a craft knife) to scrape the enamel insulation off the wire at each of the taps.

Once you've built the radio you need to connect an antenna and earth connection to the radio. The length of the antenna depends on the power of the stations you're going to be receiving and how far away they are. For more distant stations, you'll need an external antenna. Use plastic-covered stranded wire, make it as long as you can, and attach it at the far end to a tall object like a tree. Make sure it's well insulated from the support. Use a copper rod, driven into the ground, as your earth. Now, attach the crocodile clip on the flying lead to one of the taps on the coil – 50 turns would be a good starting point – put the earpiece in your ear and start tuning the variable capacitor. We trust you'll be amazed at being able to tune into one or two local broadcast stations with such a simple circuit and no battery.

Realistically, you're not going to be using a crystal set as part of your amateur radio station, however fascinating it might be. We do present a simple transceiver project in Thing 21, though, and a super-simple transmitter in Thing 22, and you could indeed use these on the air. However, both these should be considered novelty projects and don't really have a place in a mainstream amateur radio station. However, to prove that you can indeed build a simple kit that's of practical value, right here at the outset of our discussion of home construction, we're going to introduce the SWR meter. Exactly how and why you'd use it, and why it's so important as part of erecting an antenna, is covered in Thing 24. However, suffice to say it measures signals flowing from the transmitter to the antenna and those reflected signals flowing in the opposite direction. This allows the standing wave ratio (SWR), which is a measure of antenna performance, to be displayed.

You could buy an SWR meter, and they're not expensive. However, building one provides a better appreciation of how they work and could be your first opportunity to build a part of your amateur radio station. In particular, while many SWR meters are automatic devices that calculate and display the SWR from the two signals, we recommend you build a conventional manual meter that requires some adjustment each time it's used. This is a much simpler home construction project, and it really isn't difficult or cumbersome to use. Designs mostly use either one or two toroidal coils or two strip lines etched onto a PCB to sense the two opposing signals. You'll find plenty of designs online with detailed constructional notes but, to show you how simple they can be, take a look at the schematic that appears as Figure 19.2.

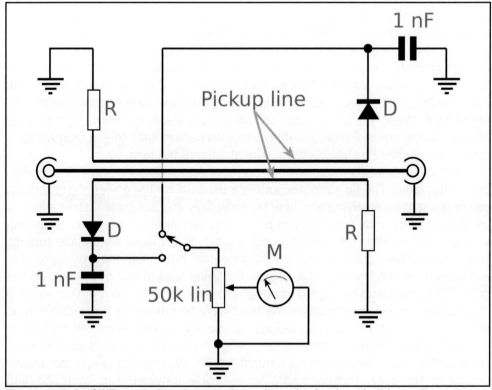

Figure 19.3. An SWR meter, like the one shown here, is simple to build as a first amateur radio DIY project, and a useful addition to your station.

Thing 20 – Communicate using Homemade Equipment

Using a radio transmitter is regulated by authorities worldwide. This is because of the potential to cause interference and thereby disrupt other radio users such as emergency services or broadcast radio and TV stations. At one time, this meant that a licence was necessary for any use of a radio transmitter, but this has been relaxed in recent years. Indeed, most of us use unlicensed transmitters on a regular basis. Included here, for example, are mobile phones, Wi-Fi access points, remote car keys, Bluetooth headphones or earpieces, and so much more. There are also licence-free radios which, superficially, resemble some amateur radio equipment, specifically PMR446 walkie-talkies and CB radios.

Despite the availability of unlicensed radio transmitting equipment, the use of such equipment is still regulated and this brings us to a unique feature of amateur radio. Equipment that can be used without a licence has to be designed so that it meets very demanding standards which are intended, primarily, to eliminate harmful interference. End users can only use suitably certified equipment, and manufacturers have to demonstrate that their equipment meets the relevant standards. So, for example, although CB radio is considered a hobby, as is amateur radio, CB operators can use only certified equipment – home construction isn't permissible.

As we turn our attention to amateur radio, we find a very different picture. The bottom line, and the relevance to our discussion of electronic construction, is that radio amateurs are allowed to build their own equipment. And it gets better, this is despite the fact that the regulations that limit the equipment are so much more relaxed for amateur radio. For example, PMR 446 handheld radios operate at 446MHz, the transmitter power is limited to 500mW, and an external antenna isn't allowed. CB radios, on the other hand, operate at 27MHz, they are limited to 4W, and there are limits on the size of the external antenna. By way

Photo 20.1. CB and PMR446 radios are small, cheap and easy to use but, for those with technical aspirations, building their own kit simply isn't an option.

of contrast, amateur radio in the UK can use 27 bands ranging from 136kHz to 248GHz, which is almost the entirety of the radio spectrum – see Thing 3. The maximum transmitted power varies from band to band, but can be as much as 1kW for Full licence holders, reducing for Intermediate and Foundation licence holders. In addition, on many bands there are no limits to the size of an external antenna. The upshot of this is that, while PMR 446 users can typically achieve a range of a few kilometres and CB operators rarely manage more than 20km, radio amateurs can span the globe.

The reason that things are so much more relaxed in the realm of amateur radio – and hence far more favourable for the user – is that, unlike PMR446 and CB, operators must obtain a licence. And you can't just buy a licence. You first need to pass an examination to prove that you have sufficient knowledge – both technical and in the realm of operating practices – to avoid causing interference to other radio users. In other words, with amateur radio, the responsibility of ensuring responsible use of the radio spectrum is transferred from the equipment manufacturer to the user. Furthermore, that user can also design and build the equipment. So, starting out in amateur radio requires a bit of work to obtain a licence, but surely that's true of most good things in life. And as you continue to read about our 50 Things you can do with Amateur Radio, we hope and expect that you'll discover that this hobby can, indeed, offer some unique and rewarding experiences.

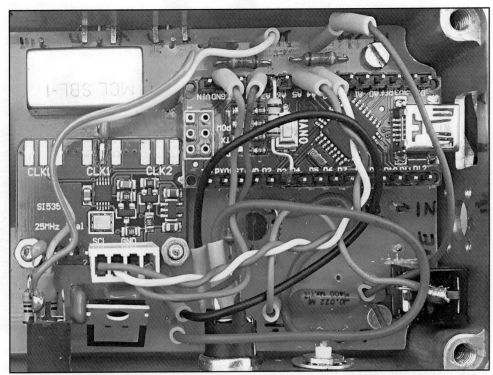

Photo 20.2. Amateur radio offers the option of building your own equipment and experiment with new techniques, as Gordon Lean, G3WJG, did in creating his SDR transceiver.

2: Construction

Thing 21 – Assemble a Kit

If you're wary of jumping straight into designing your own amateur radio gear, or even copying a published design, you might like to try a simpler solution. This half-way approach to home construction is to buy and assemble a kit, and there are plenty available for the radio amateur, some at incredibly low prices. Building from a kit means that you don't have to source all the components yourself, and this offers you several benefits. Perhaps the most important of these is that you don't have to get a PCB manufactured yourself as it'll be included as part of the kit. Second, you don't have to shop around several suppliers to find all the components you need, something that could seriously escalate the cost because of having to pay multiple shipping and handling charges. And third there will be no ambiguity about exactly which components to buy, something that might be an issue with some published designs. Some kits, especially those aimed at beginners, will provide detailed instructions, even down to describing how to identify resistors by their colour bands, and a few companies even offer technical support.

Transmitter and Transceiver Kits

As we saw in Thing 20, a unique benefit of amateur radio is that you're permitted to build your own transmitter. If you want to capitalise on this benefit, therefore, how about getting hold of a transmitter or transceiver kit? A classic example of a super-simple radio is the Pixie transceiver kit, which is available from multiple suppliers via eBay, AliExpress and the like, for as little as £1.39 plus postage. In fact, you can buy one ready built for not vastly more, but that takes away the fun of wielding your soldering iron. It operates on a single frequency in the 40m (7MHz) band, and astonishingly, it uses only two transistors and an audio amplifier IC. It also has a crystal which defines the particular frequency in the band (often 7.023MHz), plus a speaker and a handful of miscellaneous components.

It operates on CW only, so you'd need to brush up on your Morse code skills, and the transmitter power is only a few hundred milliwatts. But although reviews almost invariably say the performance isn't brilliant, and it's been described as a novelty product, it's surely superb value for money. Despite all this, we note that radio amateur Miguel Angelo Bartié, PY2OHH, in São Paulo, Brazil, has used a Pixie to make contact with stations in Argentina, Paraguay and Uruguay, the latter being at least 1,200km away.

If you want to stick with assembling kits but want to try something a bit more sophisticated than the Pixie, Walford

Photo 21.1. If you're new to assembling kits, you really can't go wrong with the Pixie.

50 Things for Radio Amateurs

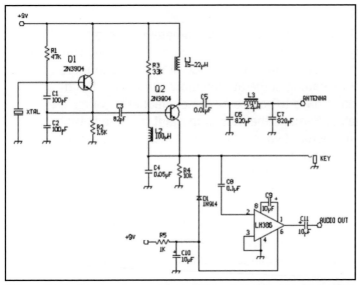

Figure 21.1. As you can see from this schematic, the Pixie 40m (7MHz) band transceiver, which is available in kit form, couldn't be much simpler

Electronics Ltd in the UK (https://walfords.wordpress.com) offers a large range of more ambitious kits for amateur radio. Included here are receivers, transmitters and transceivers for various bands, plus accessories including antenna matching units and a linear amplifier to boost the output power of a transmitter. Also, take a look at Kanga Products (https://www.kanga-products.co.uk) which has kits for Morse code trainers, practice oscillators and keyers, antenna tuning units, power and signal strength meters and more.

Miscellaneous Kits

Various other kits, while not for building amateur radio equipment, could well be of interest. First up are kits for modules such as audio amplifiers and power supplies. Using these kits could simplify the design and construction of your various projects by removing the need to design in these basic building blocks. For these and more, take a look at Kitronik (https://kitronik.co.uk), Dazzletech (https://dazzletech.co.uk), Quasar Electronics (https://quasarelectronics.co.uk) and Velleman (www.velleman.co.uk).

Building a kit could also be a cost-effective option for getting hold of some of the test equipment which, sooner or later, you're going to need as you journey into electronic construction. We note, for example, that Velleman has an audio signal generator kit and a continuity tester kit, while Mitch Electronics (https://mitchelectronics.co.uk) offers a kit for a logic probe. And for something quite different, if you want to take up the gauntlet of learning Morse code, you can get hold of Morse practice oscillator kits from Velleman and Extreme Kits (https://extkits.co.uk).

Photo 21.2. These kits, from Kanga Products, are for a Morse tutor and two CW transceivers, one a simple 2W design.

Thing 22 – Transmit using the Simplest of Equipment

You might have thought that the Pixie transceiver that we saw in Thing 21 was simple, and you'd be right. However, amateur radio gear can be even simpler and, even if you're a novice to the art of electronic construction, this might just persuade you to build something from scratch as an alternative to assembling a kit.

In particular, it's possible to build a transmitter that uses just a single transistor, as opposed to the Pixie's two transistors and one IC. To be fair, we're not comparing like with like since the Pixie is a transceiver, so if you're using our one-transistor transmitter you'll need to use a separate receiver. Assuming you have a suitable receiver at hand, though, this will actually provide an improvement since that receiver will, no doubt, offer much better performance than the Pixie. And in passing, you might like to ponder on just how astonishing it is that any electronic gear containing just a single transistor could doing something useful. After all, the smartphone in your pocket probably contains upwards of 10 billion transistors.

Our recommended design, which is a variant of a commonly published circuit, was designed by Romanian radio amateur Ciprian Popica, YO3DXE (https://dx-explorer.com). It uses just ten components, including that single transistor, and Ciprian suggests it can be built in ten to fifteen minutes. It can be used on either the 80m (3.5MHz) or 40m (7MHz) band, the exact frequency being dependent on the choice of the crystal. The output power is in the region of 700mW and reports from European operators suggest that very similar designs are capable of making contacts across the continent. The schematic of the 40m (7MHz) version is provided as **Figure 22.1**, but the list of components, which appears as Table 22.1, shows the alternative values of some components that are needed for operation on the 80m band. Note that L2 is a hand-made inductor which is wound on a T37-2 toroidal core. 18 turns of 0.35mm enamelled copper wire are required for the 40m band; 27 turns are required for the 80m (3.5MHz) band.

Ciprian recommends building the transmitter on a PCB, indeed you can buy one on his website. However, while we certainly suggest you don't build it on a breadboard, at these low RF frequencies construction on strip-board should also work well. However, it would be a good precaution to make breaks in copper strips as necessary to isolate any unused parts of the strips and thereby minimise stray capacitance. Having sol-

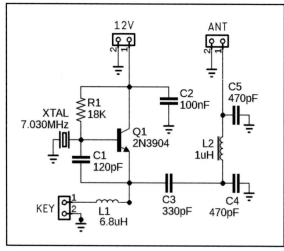

Figure 22.1. You're really not going to get much simpler than this single transistor transmitter by Ciprian Popica, YO3DXE.

Component ID	Value	Notes
R1	18k, 0.25W	
C1	120p, ceramic	
C2	100n, ceramic	
C3 (40m)	330p, ceramic	
C3 (80m)	390p, ceramic	
C4 (40m)	470p, ceramic	
C4 (80m)	820p, ceramic	
C5 (40m)	470p, ceramic	
C5 (40m)	820p, ceramic	
L1 (40m)	6.8μ, axial	
L1 (80m)	22μ, axial	
L2 (40m)	1μ, toroid	Hand-wound, see text
L2 (80m)	2.2μ, toroid	Hand-wound, see text
XTAL (40m)	40m HC49/U crystal	Choose exact value (e.g. 7.030MHz) for desired frequency
XTAL (80m)	80m HC49/U crystal	Choose exact value (e.g. 3.560MHz)
Q1	2N2222A	
12V	2 x pin connectors	
ANT	2 x pin connectors	
KEY	2 x pin connectors	

Table 22.1. Components list, showing alternative values for the 40m (7MHz) and 80m (3.5MHz) bands, for the minimalist CW transmitter.

Photo 22.1. Building this 700mW minimalist CW transistor should take you no longer than 15 minutes, especially if you use a PCB. Photo: Ciprian Popica.

dered all the components together, the only thing you need to do is to adjust L2 for maximum output while holding down the Morse key. This is done by bunching together or separating the turns on the toroid while monitoring the output with a power meter attached to the antenna connections. If you don't have a power meter here are a couple of options. First, an SWR meter, as referred to in Thing 19, can also be used to measure the signal passing to your antenna. And second, it's easy to build a low-power meter using a multimeter and a handful of components – search online for 'QRP power meter'.

2: Construction

Thing 23 – Get to Grips with Unusual Hardware

In Thing 19 we saw how RF circuitry often uses components that are not normally encountered in digital electronics, but it would be wrong to refer to them as unusual. Here, though, we're going to refer to hardware which truly is unusual, in fact you could well call it strange or even bizarre.

In the maker community, building retro projects is quite a popular activity, indeed we note that HackSpace magazine commonly publishes articles on this theme. The unusual hardware we have in mind here is just that – a blast from the past. And given that amateur radio can trace its roots back over a hundred years, such a trip down memory lane isn't inappropriate. So, how about trying your hand at building some radio equipment using vintage, and arguably obsolete, components? The most obvious component that comes into that category is the valve, otherwise known as the tube, and we discuss that later as a possible project. However, Nixie tube numeric displays, that were the predecessors to the familiar 7-segement LEDs are always popular among hackers. So test equipment with a Nixie display would be another possible project.

Introducing the Valve

Valves were the first component that allowed an electrical signal to be amplified and, in so doing, their invention heralded the birth of electronics. It was, therefore, the predecessor of the transistor and the similarities – and also the differences – can easily be appreciated.

First the differences. While the operation of transistors involves the migration of electrons (and holes) through solid semiconductor materials, valves work via electrons flowing through free space. That space must be a vacuum, which is why valves are housed in glass cylinders. Furthermore, electrons are only released when the valve's cathode is heated to a high temperature. So there's a filament inside the cathode, which heats up and glows red when a low voltage is passed through it. And finally, because an appreciable flow of electrons through a vacuum only occurs when there's a high potential difference, valves might use higher voltages than transistors.

Next the similarities. A valve is an amplifier and, in that respect, it's similar to a transistor. It's actually more similar to an FET (field effect transistor) but, since bipolar PNP transistors are more familiar, we'll compare a valve with this component. So, we can consider the cathode to

Photo 23.1. Relive the days of valve radios, fragile glass enclosures, warm red glow and all. Photo: Shane Gorski.

61

be similar to the emitter on a transistor, the anode is akin to the collector, and the grid is equivalent to the transistor's base. More specifically, we're referring here to a triode valve which has just one grid, which is referred to as a control grid. Other valves, such as tetrodes and pentodes, have additional grids, but these are just to make their operation more linear and are usually just connected to various supplies.

Valve-based Project

If you're inspired to build something that's based around a valve, our recommendation is to create some simple radio equipment. One option is to add a valve-based audio amplifier to the crystal radio we looked at in Thing 19. This would allow it to drive a speaker instead of headphones or an earpiece. However, our more specific suggestion is a single frequency transmitter like the one we saw in Thing 22, but using a valve instead of a transistor. We're not going to provide full construction details. However, we will paint a picture of how you might proceed en route to transmitting with archaic equipment while bathed in the warm glow of a valve's characteristic red light. And if it's good enough for the retro-minded audiophile enthusiast, it's good enough for the radio amateur.

Circuits are by no means as easy to find as for single transistor transmitters, but you will find them online and, with a bit of perseverance, you should find something that meets your needs. However, some can be as simple as the one shown in **Figure 23.1**, which is a variant of a design first published back in the mists of time. The valve's heater is the semi-circle at the bottom of the symbol, the cathode is just above it, the grid is the dotted line, and the anode is at the top. This circuit is based around the 955 valve, which is an obsolete design referred to as an acorn valve, because of

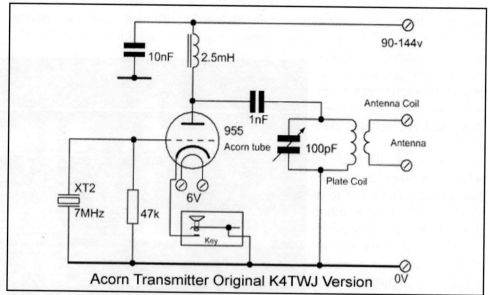

Figure 23.1. Single valve transmitters, like this vintage design, can be almost as simple as transmitters using just one transistor.

its rather quirky shape. You can still get hold of them, often second hand, but if you do decide to use a variant of this circuit, it might be better to swap the valve for something more readily available. However, that might involve changing the values of some of the other components. Here are some tips on constructing valve circuits.

Valves are not nearly as widely available as more modern components, but a few manufacturers make them, including JJ Electronic in Slovakia (**Photo 23.2**), and some mainstream component suppliers sell them. Some specialist suppliers also stock valves, often ones that are brand new but have been sitting in warehouses ever since they were manufactured decades ago. However, they can be expensive, so do shop around.

Like ICs, valves plug into sockets, usually called bases, through which they're wired to the rest of the circuit, so you'll need a base to fit your valve. Normal practice was for valve circuits to be built on an aluminium chassis, which was basically a box, often with no bottom surface. The valve bases were mounted into holes on the top surface of the chassis, which meant that the bases' pins, to which soldered joints are made, would be underneath the top surface. So, the passive components were fitted underneath the chassis, and any user controls would be on the front of the chassis. If you really hanker after that retro look, you might want to replicate this method of construction.

Photo 23.2. Valves are still manufactured by a few companies. Alternatively, unused old valves are available.

Photo 23.3. Valve circuits were built using a very different method of construction from today's PCB-based techniques. Photo: Joe Haupt.

When you start working with valves, you shouldn't lose sight of the fact that they require a supply in excess of 100V, even up to 1,000V or more for high-powered transmitters. So, make sure you take every safety precaution in building and debugging your circuits.

If all of this sounds just too much trouble, here's an alternative solution that's much easier. We note that kits of parts for a valve-based 7MHz band CW transmitter are available from several suppliers, mostly via AliExpress. Prices start at just £7.93 including delivery. These kits don't promote construction on an aluminium chassis. Instead, they tend to be of a hybrid nature in having the valve mounted on a PCB. Beware of the kits for AM medium wave transmitters, which also abound, though. These won't allow you to transmit on any amateur band, and using them would be illegal.

Thing 24 – Experiment with DIY Antennas

You might expect that a good way to increase the range of your transmitted signal is to increase your transmitter's output power. And while that would certainly work, there's arguably a better solution. That solution is to improve the antenna, and here's why. Doubling the transmitter power increases the radiated signal by 3dB. This means that to get a 6dB improvement requires four times the power, while a 9dB increase requires the power to be increased by a factor of eight. Clearly the law of diminishing returns soon makes itself felt. By way of contrast, it's often easier to get these levels of improvement by using a better antenna, especially at higher frequencies. What's more, while increasing the transmitter power only improves your transmitted signal, improving the antenna also provides benefits for receiving weak signals. For this reason, we feel entirely justified in making this topic the longest one on the subject of home construction.

The standard type of antenna, against which other antennas are often compared, is called a dipole and we'll investigate it a bit later. Although a dipole exhibits directionality – i.e. it transmits more signal in some directions than others – other antennas are even more directional and this is the key to their improved performance. To give a quite impressive example, a ten-element Yagi antenna for the 70cm (430MHz) band is reasonably compact yet its gain – that is the improvement over a dipole – is in the region of 13dB. In other words, it has the same effect as increasing the transmitted power by about 16 times. Admittedly gain figures will rarely be this high at lower frequencies because of size constraints, but even a much more modest gain would be well worthwhile. So, experimenting with antennas will often be a good investment.

Building an antenna is certainly a form of home construction, although the skills required are rather different from those needed in building a transmitter or receiver. For example, soldering is only a minor requirement and, instead, you're more likely to find yourself up a ladder providing fixings to your house or perhaps a tree. As your skills grow, you might find yourself installing huge directional and rotatable HF antennas on lattice towers.

Photo 24.1. In time you might aspire to building a huge 'antenna farm' like this, but for now we've got some easier options for you to consider. Photo: Евгений Катышев.

2: Construction

As an introduction, though, we're going to look at a few simple antennas for use in the shortwave bands and one for VHF or UHF which, nevertheless, can offer good performance.

Ancillary Equipment

Before looking at any particular types of antenna, we need to mention a couple of pieces of equipment that you'll need to get the most out of your antennas. In particular, we're going to take a look at the antenna tuning unit (ATU) and the SWR meter.

As its name suggests, an antenna tuning unit 'tunes' an antenna. In other words, it adjusts the impedance at a given frequency so that it matches the impendence of the transmitter. Unless an antenna is matched in this way, some of the signal sent to the antenna is reflected back to the transmitter. This results in a high value of the so-called standing wave ratio (SWR). This, in turn, reduces the amount of signal actually transmitted, and can result in damage to the transmitter. A mismatched antenna also has a negative effect on the received signal. An ATU is essential with un-tuned antennas like the long wire, which is the first type of antenna that we'll introduce. However, it might also be beneficial with some tuned antennas – that's an antenna designed for a particular frequency – because the impedance might not be ideal. ATUs comprise a network of variable capacitor(s) and an inductor with switch-selectable taps and, at one time, radio amateurs would invariably build their own. With variable capacitor prices now being so high, this is a less attractive option today. So, unless your transceiver has a built-in ATU, you might prefer to buy either a manual or automatic ATU.

Having already mentioned the SWR as a measure of how well your antenna is matched to the transmitter, we should introduce the SWR meter which allows you to measure this metric. We referred to an SWR meter in Thing 19 as a possible constructional project, and here we look at why and how you'd use one. SWR is defined as $(1 + (R/F)) / (1 - (R/F))$ where R is the magnitude of the reflected signal strength and F is the magnitude of the forward signal. This means that if there's no reflected power the SWR is 1:1, and if all the signal is reflected the SWR is ∞:1. Needless to say, 1:1 is ideal, and it's generally accepted that anything below 3:1 is acceptable. You'll use the SWR meter when you're

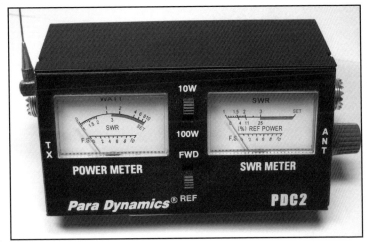

Photo 24.2 SWR.jpg. An SWR meter, which can either be commercial or a DIY project, is an essential tool in antenna tuning. Photo: Vaughan Weather.

'tuning' a long wire antenna with an ATU to see when you've adjusted it correctly. And, in the case of a tuned antenna like a dipole, an SWR meter is an important tool in cutting the wire to the correct length.

Long Wire Antennas

The simplest possible antenna, and often the first that radio enthusiasts will use, is often referred to as a long wire antenna. Actually, the word 'long' isn't always accurate – especially if you have a small garden – so the phrase random wire antenna might be more appropriate. The length is usually chosen to fit the available space and, as a result, it probably won't offer ideal performance on any particular frequency. If your first foray into amateur radio is going to involve listening rather than transmitting (see Thing 1) you might have acceptable performance across a range of frequencies. However, you'll get better results if you tune the antenna to whatever band you're listening to, using an ATU. And when you start transmitting, tuning your long wire is pretty much essential.

A long wire is exactly what the name suggests – a length of wire that plugs into the inner part of the antenna socket – although the practicalities are a bit more involved. It really needs to be as high as possible, so you'll probably need to attach it to a tall object such as a tree at the far end, and to your house at the end that plugs into your equipment. Use insulated stranded wire and, as an extra precaution, use insulators and cord at the ends to keep it insulated from any grounded supports. This is clarified in **Figure 24.1**. A good ground connection to the outer part of the coaxial antenna socket – ideally a buried length of wire or earth rods driven into the ground – is also recommended.

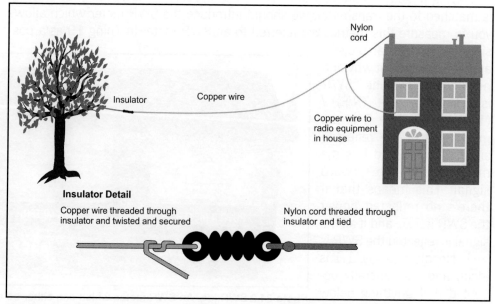

Figure 24.1. **A long wire is an excellent choice as a first antenna, especially since it can be tuned to any HF band.**

2: Construction

Dipoles
A much better option than a long wire antenna, at least at the higher frequency shortwave bands, is a dipole and, in particular, a horizontal half-wave dipole. This is tuned to a particular frequency, although you can consider it as useable over a complete band if you design it for the middle of the band by cutting it accordingly. The tuned nature of a dipole is both a blessing and a curse. On the positive side, it doesn't need an antenna tuning unit; on the negative side, you'd need one for each band, although it's possible to wire several dipoles in parallel on the same feeder. Alternatively, it's not uncommon to use a long wire on 160m (1.8MHz), 80m (3.5MHz) and perhaps 40m (7MHz), and dipoles for the higher frequency bands. Generally speaking, it's harder to accommodate a dipole than a long wire of the same length because it's centre-fed rather than end-fed. It's more problematic, therefore, unless you have space at both sides of your house.

The arrangement of a horizontal dipole is shown as **Figure 24.2**. Attaching the ends of the antenna to its supports is carried out in the same way as the far end of a long wire. Where it differs, however, is at the centre, where the antenna is cut into two halves. First these two halves are soldered onto the inner and outer conductors of the coaxial cable which acts as the feeder, connecting the antenna to the radio equipment. And second, a support might be required in the centre,

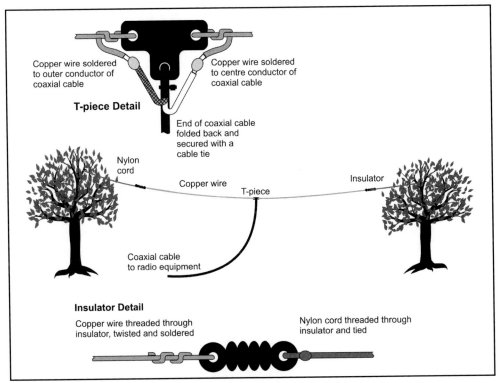

Figure 24.2. A dipole is a good choice but it does require two supports, one either side of the feed point.

67

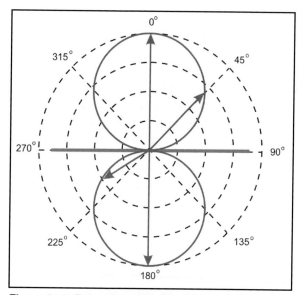

Figure 24.3. Polar diagrams, like this one for a dipole, illustrate the directional properties of an antenna.

to keep it well away from the house or any other obstruction. Actually, it would be possible to route a dipole in the same sort of way as a long wire, i.e. with one end at the house. The disadvantage is that you then need a longer feeder which can cause losses and, furthermore, that feeder should be routed along a path that isn't in close proximity to the antenna.

The other vitally important thing you need to know is how long the antenna should be to ensure it's tuned to a particular frequency. A dipole should be half a wavelength long, split into two lengths of a quarter of a wavelength. It's easy to work out, though, using the formula L = 143 / f, where L is the total length of the dipole in metres, and f is the frequency in MHz. So, for example, if you want a 20m (14MHz) band dipole centred on 14.200MHz, the total length should be 10.07m. Because this can be influenced by several factors, it's common to cut the wire somewhat longer than calculated, and trim it until ideal performance is obtained, as measured using an SWR meter. Note also that a dipole is directional, transmitting or receiving maximum signal broadside to the antenna and, in theory, nothing from the ends. If you have a choice, therefore, this might dictate how you orient it. Directionality of antennas is illustrated using polar diagrams, like the one for a dipole which appears as **Figure 24.3**. The dipole antenna is shown here as the horizontal line while the two circles represent the amount of signal in any direction. As you can see from the lengths of arrows - which don't normally appear on polar diagrams - the maximum signal is perpendicular to the antenna and drops off towards the ends

Quarter-wave Verticals
One other simple type of antenna that you might like to experiment with is a quarter-wave vertical – see **Figure 24.4**. Its length is half that of a half-wave dipole and it can either be mounted directly on the ground, but insulated from it, or on some form of elevated platform, even the roof of a building. For shorter antennas – that is for a shorter wavelength and hence a higher frequency band – the antenna could be self-supporting but longer ones will need guys. The antenna can be made from aluminium tubes – perhaps reducing in diameter towards the top – fixed securely to the ground or elevated support. Alternatively, you could use an insulting pole of some sort, with a wire securely attached to it – fishing rods have been used for this purpose. This sort of vertical antenna connects to the inner conductor of the feeder while the outer conductor should be connected to an earth. If the antenna

is mounted on the ground the earth connection could be one or more spikes driven into the ground. However, much improved performance is achieved using a radial array of quarter-wave wires resting on the ground or buried just below the surface. If it's mounted above ground level, the only option is to use the radial wire approach, not buried of course, to create an artificial ground. In this case you might be limited to four radials. It's also easier in this case if the radials are angled downwards, perhaps at 45°, a configuration that's often called a 'ground plane antenna'.

A quarter-wave vertical antenna is omni-directional when viewed from above. However, it does exhibit a degree of directionality in the vertical plane and this gives rise to a 3dB gain compared to a dipole.

Yagi Antennas

The name might sound strange – in fact this type of antenna is named after the Japanese electrical engineer Hidetsugu Yagi – but the Yagi is one of the most common types of beam antennas. A beam is a type of antenna that achieves a high gain by focussing the transmitted signal in a particular direction. Its polar diagram is similar to the one we saw for a dipole (Figure 24.3), except that one of the two circular lobes is narrowed but lengthened while the other one is reduced in size.

A Yagi antenna is based on a dipole but has additional parasitic elements which are parallel to the dipole. These parasitic elements are in close proximity to the dipole but have no connection to it. As a minimum, there will be one parasitic element called a reflector which is somewhat longer than the dipole's half-wave and separated from it by about a quarter of a wave. Usually there will also be one or more elements called directors, slightly smaller than the dipole, on the other side. A two-element beam has a gain of about 5dB compared to a dipole, and this increases

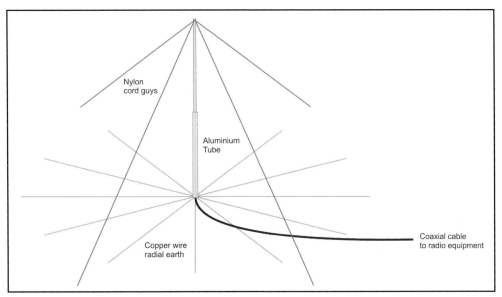

Figure 24.4. A quarter-wave vertical antenna is a good choice, but for longer wavelengths you'll need to guy it.

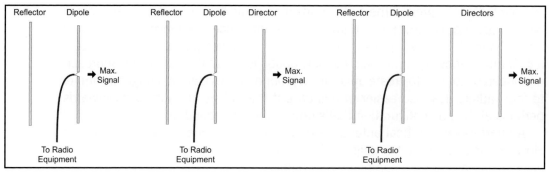

Figure 24.5. Yagi antennas can offer a significant gain over a dipole, the gain increasing with the number of elements.

to 9dB for three elements. Each additional director improves the gain, although the law of diminishing returns applies. Typical arrangements are shown in **Figure 24.5** which compares a two-element, a three-element and a four-element Yagi.

Moving to the practicalities, it soon becomes clear than Yagis will be large in physical size on the HF bands. When we consider that these antennas need to be rotatable, it is also evident that there's a limit to which bands they can be designed for. Yagi antennas, albeit often with just three elements, are used down to the 20m band or very occasionally the 40m band, but they're massive as you can see half way down the tower in **Photo 24.1**. In the 2m (144MHz) and 70cm (430MHz) bands, though, they're much more practical, and multi-element, high-gain Yagi antennas are common. You'd be advised to follow a published design to determine the element lengths and their spacing, but here's the gist of how to construct one. The elements are held together using an aluminium tube referred to as the boom. The parasitic elements, which are made from smaller diameter tube, can be attached directly to it without insulation, although the driven element, i.e. the dipole, must be insulated from the boom. This is then attached to a mast which can be rotated either by hand or, ideally, with a motor-driven rotator.

Photo 24.3. This Yagi antenna, which uses a folded dipole rather than a regular one, and unusually two reflectors, will achieve a gain of around 15dB. Photo: Ks.mini.

Thing 25 – Setup a Mobile Radio Station

Your main radio station will probably be located in your home, but setting up a mobile amateur radio station in your car is also a popular and challenging project. Commonly, a mobile station will operate on 2m (144MHz) or 70cm (430MHz) for local coverage, often via repeaters. However, amateurs have made contact all over the world from their cars using high-power HF radio.

We consider this as a constructional project, in the broadest sense of the phrase, even if you base your mobile station on a commercial transceiver. This is a major topic which warrants some serious attention, so our aim here is just to sow a seed of inspiration. And if you decide to go down this route, you really ought to seek out expert guidance. After all, if you don't take care you could seriously impact the resale value of your car, and there could also be safety concerns.

So, what do you need to do to get up-and-running on the road? First you need a method of mounting your transceiver so it's easily accessible while you're driving. Then you need to route a power source to it. This should be wired directly to the battery so there's no possibility of impacting the car's many electrical systems, and should be suitably fused. And last, but by no means least, you need to mount an antenna – commonly a vertical – on your vehicle, and you need to route its coaxial feeder to your transceiver. This will require you to give serious thought to optimising performance while minimising the impact to your car. In reality, this will probably involve making compromises. For example, you'll undoubtedly get best performance by mounting the antenna on the roof, but drilling a hole in the roof would be hard to justify. And even if you adopt the alternative magnetic mount approach, reports suggest that scratches and scuffs could well result.

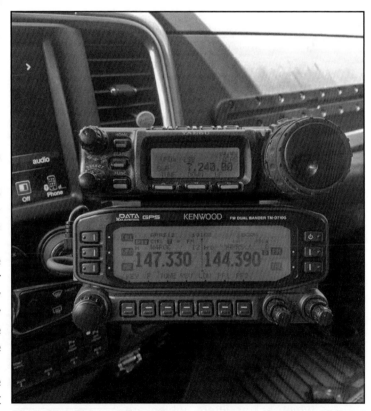

Photo 25.1. This mobile station can operate on shortwave, VLF and UHF, despite the modest dimensions. Photo: John Fury

Thing 26 – Help Amateurs by Building a Repeater

Most of the construction ideas we've look at so far are suitable for beginners. Our next suggestion of what you can do with amateur radio is much more ambitious. While that means that novices aren't likely to undertake such a project in isolation, building and operating a repeater is often a group activity in which you might well be able to play a role.

Repeater Circuitry

As we saw in Thing 7, that amateur radio repeaters are used to increase the range of low-power stations, often handheld or in-car mobile stations operating on the VHF or UHF bands. Repeaters achieve this by receiving a signal on one frequency and re-transmitting it simultaneously on another frequency, commonly in the same amateur band. **Figure 26.1** is a simplified block diagram of a typical repeater. As you can see, there are two elements which stand out as ones that aren't normally found in amateur radio stations.

At the heart of any repeater is a control unit which, typically, has several functions. It causes the repeater to periodically identify itself in Morse code, it controls access to the repeater by use of an audio tone, it enforces the maximum time a single station can access the repeater, and it provides a means of closing down the repeater remotely. More advanced features could include management of voice announcements and, interfacing to the internet. The other unusual element of a repeater is its cavity duplexer. This allows the repeater to transmit and receive at the same time, using the same antenna, on very close frequencies. This vital element protects the receiver from the catastrophic damage that would otherwise occur. The cavity duplexer contains tuned circuits which act as filters, which are very selective and can handle the moderately-high transmitter

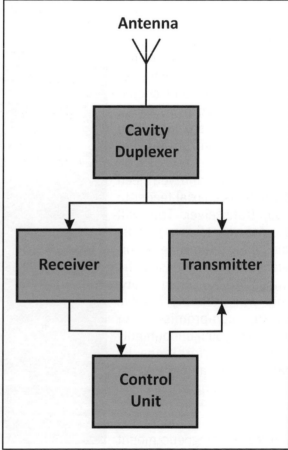

Figure 26.1. As you can see from this block diagram, repeaters use a couple of building blocks that aren't found in ordinary amateur radio equipment.

2: Construction

power. Building and constructing the duplexer requires skills which, to the layperson, seem to have more in common with plumbing or mechanical engineering than electronics.

Skills Required in a Repeater Project

Our discussion of the block diagram will have identified a couple of areas of electronic design that you might be able to contribute to a repeater project. And although a decision might be made to use a commercial or open source controller, such devices still require programming to meet the needs of a particular repeater. However, there's much more. At the outset of the project, a repeater group might consider modelling the coverage of prospective sites. So, if you fancy getting to grips with modelling software, this could be for you. Not exactly construction, but another important role in any repeater project is submitting and managing the licence application. And, in the latter stages of the project, the antenna needs to sourced, and all the gear needs to be housed, interconnected and installed. Next up is testing and, in due course, the big switch on. Even this isn't the end of the project, though. You might also like to be involved in the ongoing operation, maintenance and possible future upgrade of the repeater.

Photo 26.1. A controller is at the heart of any repeater and options range from using a commercial unit through to building a published circuit to designing one from scratch. Photo: Sterling Mann.

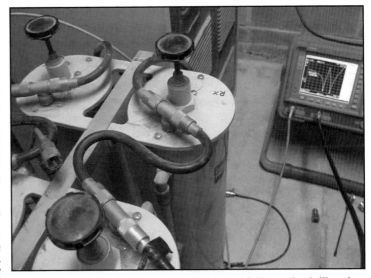

Photo 26.2. These large metal cylinders might not look like electronic circuits but they comprise the cavity duplexer which is essential to repeater operation. Photo: Sterling Mann. .

Thing 27 – Upgrade your Coding Skills

Many of the concepts and components that are used in conventional radio equipment are largely unfamiliar to those whose electronics experience has concerned computer interfacing, as we saw in Thing 19. So it might be somewhat surprising that we're now turning to the subject of using your coding skills to create radio equipment. But while much radio equipment is still packed full of analogue electronics, things are changing. This brings us to the topic of software designed radio (SDR) and how this could be a vehicle for you to employ your coding skills.

SDR Overview

Although software defined transmitters certainly exist, the SDR approach has mostly been applied to receivers – at least for DIY projects – because the benefits are more obvious. You could buy an SDR receiver or transceiver off the shelf. It might take the form of a box that attaches to your PC to provide control or, alternatively, it might not look dissimilar to conventional equipment. However, a more cost-effective solution, and one that better exercises your technical skills, is to create one yourself. One possible approach to SDR receiver design is to feed the antenna signal to an analogue to digital converter (ADC), perhaps via an RF amplifier, and then carry out virtually all the functionality of a conventional receiver using software. While this is often considered the ideal solution, in that it offloads as much as possible of the functionality to the software, it does have some disadvantages. First, the ADC has to convert a large portion of the RF spectrum which means it has to sample at a high frequency and is, therefore, expensive. Secondly, processing the RF signal directly is hugely processor-intensive which means that much of the software has to reside on esoteric high-speed hardware such as a field programmable gate array (FPGA).

Photo 27.1. Commercial rigs like the Icom IC-905 are based on SDR technology, but this approach also lends itself to DIY experimentation.

2: Construction

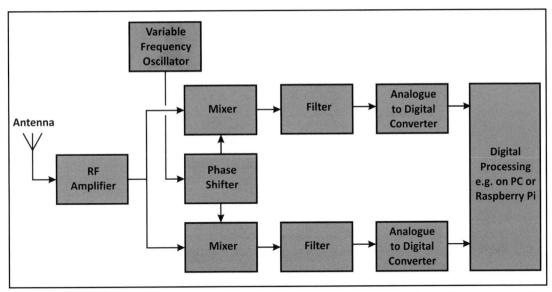

Figure 27.1. Block diagram of an approach to SDR receiver design commonly used in amateur radio.

Although radio amateurs have designed receivers using this type of architecture, an alternative approach, that's illustrated as a simplified block diagram in **Figure 27.1**, is commonly used instead.

The major advantage of this alternative architecture is that, rather than directly converting the signal from the antenna to the digital domain, a much narrower band of frequencies is passed through to the ADC. This both reduces the sampling rate and cost of the ADC, and it allows the digital processing to be carried out on simpler hardware such a PC or Raspberry Pi. This is achieved by mixing the antenna signal with the output of an oscillator, which is tuned to the frequency of interest, and filtering the output of the mixer. In fact, for reasons we'll gloss over here, but essential for the software which follows, two ADCs are used to provide two phase-shifted signals called 'I' and 'Q'.

RTL-SDR Dongle

A popular approach to providing the hardware element of this sort of SDR receiver is to use an RTL-SDR dongle. This type of USB-attached device, available cheaply from several suppliers, was originally intended for receiving digital TV signals. Do read up on these devices before buying one, though, to make sure that you obtain one that is suitable for amateur radio applications. In other words, you need one that is capable of tuning down to cover the amateur HF bands. Once you have an RTL-SDR dongle, all you need is suitable software. This provides the functions of tuning, filtering, demodulation and decoding the signal, which are necessary to turn it into an SDR receiver. Certainly, you can get free software to do that but, if you're hankering after engaging in a DIY project, you have several options. First of all, if you want to get a feel for SDR

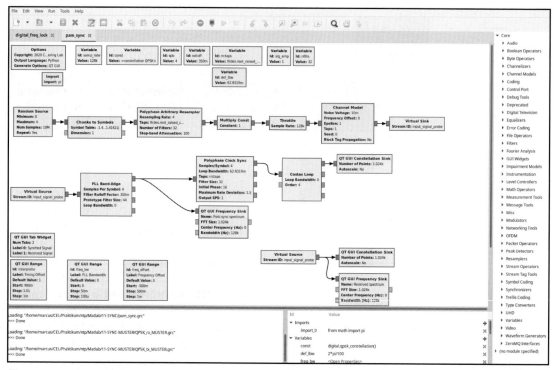

Figure 27.2. GNU Radio offers an interesting alternative to conventional coding for SDR applications. Image: Marcusmueller ettus.

software without starting a project from scratch, you could try modifying open source code. Going beyond this, though, you'll probably think of writing your own software, in all probability using your favourite programming language. Using an SDR toolkit will simplify matters, and Python is probably the best supported in this respect. However, an alternative to a conventional programming language is to use a block-based approach. Here, instead of stringing together instructions to execute in sequence, the functionality is defined by wiring together functional blocks on a screen. An especially interesting example, in that it's intended primarily for designing radio-related code, is GNU Radio, and the good news is that it's free.

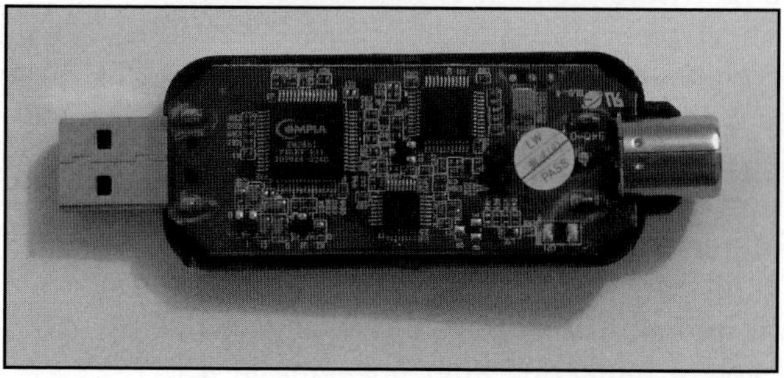

Turning our attention to the computing platform, you're free, to a large extent, to use the hardware and operating system of your choice. PCs running Windows or Linux are popular choices, as is the Raspberry Pi. The latter is particularly useful, of course, in that it allows you to build a go-anywhere SDR receiver.

Photo 27.2. RTL-SDR dongles offer a low-cost entry to DIY SDR receiver design. Photo: Wilfredor.

Thing 28 - Use our Raspberry Pi in New Ways

For our next thing in our roundup of how home construction plays an important role in amateur radio we're sticking with software. However, while SDR projects, like those we saw in Thing 27, are often entirely software related, we're now looking at projects that could have a hardware design element too. This takes us to the subject of that most popular of all single-board computers, the Raspberry Pi (RPi). What's more, it's a platform that has launched many an amateur radio project. So, if you consider yourself a Raspberry Pi enthusiast, this is for you. Alternatively, if you've never delved into the Raspberry Pi, amateur radio could provide the ideal opportunity to discover it.

Introducing Raspberry Pi

If you're a newcomer to the Raspberry Pi, here's a brief introduction. These computers take the form of a single circuit board and various models are available. Most are approximately the size of a credit card, although there are some much smaller ones. Excluding the Pi Pico, which can't run an operating system, they cost around £15-£75. Models mostly differ in the processor type, the amount of memory, and the number of ports such as USB. Add a power supply, keyboard, mouse and monitor, and you have a PC running an operating system that's normally a Linux distribution. However, the reason they're so popular among those with an interest in electronics is that they have a GPIO (general purpose input/output) port which

Photo 28.1. Despite being several years old, the Raspberry Pi 3 variants remain popular and cost from £25. Photo: BlueBreezeWiki.

allows them to be connected to external hardware. This can take the form of a HAT, which is an add-on daughter board that plugs onto the RPi's GPIO pins, or it could be an off-board circuit. While so many different HATs are commercially available, adding your own hardware is entirely possible and the basis of many a RPi-based project.

Raspberry Pi Projects

Many of the ways in which radio amateurs have utilised the Raspberry Pi are applications which could, alternatively, have been run on a PC. In some cases this choice of platform is purely to get some experience of using the RPi, but in other cases the rationale was to achieve a more portable solution. This could be useful,

Photo 28.2. Add-on boards, referred to as HATs, allow external circuitry to be connected to a Raspberry Pi via its GPIO header. Photo: Multicherry.

for example, when operating in a tent during a Field Day contest – see Thing 30. For many of these applications you can use readily available software, so this type of project might be suitable if you're new to the RPi before you start hacking your own code. Included here are SDRs (see Thing 27) and encoders/decoders for state-of-the-art digital communications modes (Thing 10). We note that RPi boards have also been used as repeater control boards (Thing 26).

Typically, these types of projects require a Raspberry Pi to be connected to external hardware such as an LCD panel and a battery power supply, and a simple interface to a transceiver. All of this needs to be housed in a suitable enclosure. But what if you want to go beyond this and perhaps engage in a project that involves some coding? Here are a few other ways in which radio amateurs have put a RPi to work.

An automatic Morse keyer (see Thing 9) is fairly easy to build using a handful of logic ICs. Even so, implementing one using a Raspberry Pi would be an interesting and educational first project. What's more, you could easily extend it to allow messages to be recorded and played back, features that aren't as easy to implement using hardware.

Using a Raspberry Pi and some dedicated hardware has allowed amateurs to create WSPR beacons (see Thing 11) that don't tie up a PC or the main transceiver.

RPi-based clocks are popular projects. Needless to say, amateur radio clocks show more than just the time and you could choose what you want to display. Satellite positions on a map, weather, solar flux, and DX spot information are just a few options.

A more ambitious project, since it also requires mechanical engineering skills, is to create an automated antenna rotator – adjusting the azimuth and elevation – for use with satellite communication.

You can get much more detail from Raspberry Pi Explained, by Mike Richards, G4WNC, that's available from www.rsgbshop.org.

2: Construction

Thing 29 – Attend a Rally

In Thing 19, as part of our consideration of how you can expand your construction skills, we looked at some components that are important in radio circuits, but are otherwise largely unfamiliar. Unfamiliarity isn't a problem in itself – after all, you can soon learn – but it rather goes hand in hand with scarcity. Indeed, the inductors and variable capacitors we discussed in Thing 19 are not commonly available from mainstream electronic component suppliers and tend to be expensive. Valves, aka tubes (see Thing 23) are also difficult to get hold of and are expensive, and other components can also by tricky. And the appropriate wires, feeders, plugs, insulators, masts and the like, that are needed to install DIY antennas (Thing 24), are only available from more-specialist suppliers. This begs the question of how you're going to get the parts you'll need for your radio-related construction projects.

Certainly you'll be able to find all these rare components online, but there's a drawback other than potential cost. Except for specialist amateur radio suppliers – and you certainly shouldn't discount those – many of the suppliers are on eBay and AliExpress. And while some of these merchants are locally based and can provide advice, this could be trickier with those based in China. If you're an old hand at radio-related electronic construction this probably won't be a problem. However, if you're new to amateur radio, or returning to it after some time, discussing what you need with suppliers is surely a good idea. Something to consider, therefore, is attending an amateur radio rally, as they tend to be called in the UK. Alternatively, they might be called exhibitions, conventions or hamfests, especially in other countries.

Photo 29.1. There is usually lots of equipment to see and buy at an amateur radio rally.

What to Expect

A rally is first and foremost an amateur radio trade fair, albeit one that hosts stands by non-profit clubs and societies in addition to commercial organisations. Of particular interest to those looking for components and the like, though, are the equipment and component dealers. Typically you'll find suppliers of new and second-hand equipment such as transceivers, receivers and antennas; components, including mainstream, computer-related and RF-specific; materials for building antennas; plus books and more. What's more, these stands will be manned by knowledgeable people, so advice won't be in short supply. And there's something to be said for actually seeing stuff before buying it.

In addition to traders, other exhibitors include amateur radio clubs plus national radio societies, often representing special interests within amateur radio. Many rallies also offer an opportunity for all attendees to sell their unwanted gear in various ways. And while it's very much the exception rather than the rule, a few rallies host a programme of talks. Finally, while rallies don't usually facilitate it specifically, radio amateurs often appreciate the opportunities that rallies provide for meeting like-minded individuals.

Figure 29.1. The National Hamfest is the UK's largest amateur radio event.

While most rallies are organised by local amateur radio clubs, a few are regional or national events, so these tend to be larger and here we should mention the UK's National Hamfest. For the ultimate, however, you'll probably need to venture overseas. Of particular note are the Dayton Hamvention in the US, the HamRadio exhibition in Friedrichshafen, Germany and the Tokyo Ham Fair. .

Photo 29.1. Some rallies have outside flea markets or are car boot style events.

3

Outdoor Pursuits

Introduction

At first sight, amateur radio might sound like a purely technical activity that is, by and large, carried out from the comfort of your own home. So, if you like to get out and about in the great outdoors, perhaps hiking in the hills or just enjoying the countryside, then taking up a new activity that will keep you indoors might not sound too attractive a proposition. But it doesn't have to be like that. Indeed, amateur radio offers you the opportunity to engage in a technical pursuit while also enjoying an outdoor lifestyle and, in so doing, keep yourself healthy and fit. So, in this chapter, we shed some light on ways in which you can enjoy the unusual combination of cerebral and physical activity.

First of all, we'll investigate the concept of a field day. These events, which take place around the world, provide the opportunity for amateur radio clubs to compete with one another as they set up and operate a temporary station from a remote location. Going one step further – or actually several thousand miles further in most cases – we get to grips with DXpeditions. Such events take place in far flung quarters of the globe that don't normally

50 Things for Radio Amateurs

enjoy amateur radio activity. As such, the station set up in this little-known location will be much in demand among the global amateur radio community.

DXpeditions are certainly for travellers and explorers and, while they aren't always outdoor affairs, they certainly can be as our next "thing" illustrates quite vividly. Admittedly, this certainly isn't for everyone since it requires very particular skills, but we trust it'll be inspiring. What we're talking about is a recent DXpedition to Rockall, that tiny rock outcrop in the North Atlantic. And while you're not necessarily going to follow in the footsteps of this DXpedition, it might encourage you to engage in a similarly unusual outdoor activity.

Less demanding, but certainly with the potential to give you a good workout, is the opportunity to operate a portable station from the summits of hills and mountains. Called Summits on the Air, this programme encourages amateurs to operate from mountain tops, the rarer the better. This can be competitive if you want, with awards available for those setting up stations and those making contact with them. Radio when you're out and about on foot is also the theme of our next "thing". Described as a combination of radio direction finding and orienteering, amateur radio direction finding involves competitors completing a course as quickly as possible while finding hidden radio beacons en route.

Providing a public service is part of the amateur radio ethos and our next two things pick up on this theme. First of all we take a look at the work of radio amateurs in providing a communication service for the community. For example, this might be to support groups involved in the organisation of outdoor events such as fell races. These groups also offer their services in times of emergency or infrastructure damage, and natural disasters such as storms, floods and even earthquakes in some parts of the world. And finally, we see how radio amateurs have played their part in supporting cave rescue teams. By developing techniques and equipment capable of communicating between the surface and underground, they've permitted rescue controllers to remain in contact with the underground teams. Lives have probably been saved as a result.

3: Outdoor Pursuits

Thing 30 – Operate During a Field Day

Amateur radio has a competitive element as we saw in Thing 13 when we looked at contests. Ordinary contests generally don't restrict where competing amateurs locate their station, so most commonly, they operate from their usual home location. This is where field days differ, because these contests are specifically for portable stations.

What's Involved?

Commonly, entrants are amateur radio clubs since that simplifies the setting up, operating and supporting of the station. After all, there's a lot more to do than talking into a microphone. Other essential activities include carrying bulky equipment including batteries or generators, installing antennas, erecting tents, setting up the equipment, and providing catering services. Rules differ, but here's how the RSGB currently define the requirements for a portable station. "Portable stations must be powered by on-site generators or batteries or renewable energy source and must not be set up in a permanent building or structure."

A definition like this wouldn't prohibit a station being set up close to home, or even outside a member's house with a large garden. However, entrants tend to enter into the spirit of a field day by locating their station somewhere more remote. And there are benefits on offer too. Most notably, choosing a location at a high elevation, especially with a largely uninterrupted view to the horizons, will very much improve a station's performance. Some clubs also take advantage of setting up the station somewhere remote – so club members will remain on site for the entire event – to also make it a social event.

Photo 30.1. Field Day contents allow you to hone your skills in portable operation. Photo: Gerd R. Sapper, DJ4KW.

50 Things for Radio Amateurs

Photo 30.2. Using portable power sources and suitably packaging your equipment are all part of the Field Day experience. Photo: Greg Heartsfield.

Field day stations are often set up near to a road, so you might feel that it barely counts as an outdoor activity, even if you do spend the night under canvas. If you hanker after the rigours of the great outdoors, though, we suggest that a field day could give you more of a work out than you might expect. You might be well accustomed to hiking several kilometres up a mountain, but don't underestimate how much difference it makes to carry a 12V battery weighing 20kg. Looking after an antenna mounted on a portable tower might also test your outdoor skills and perseverance, especially if a storm arrives during the night. These unusual forms of exercise might not compete with taking in the views while walking in the hills, we admit. Even so, your efforts and commitment will surely be appreciated by the rest of your team.

Backpackers' Contests

Strictly speaking not a field day but, on a very similar theme, the RSGB organises an annual 144MHz Backpackers' Contest. So, if a field day's high powers and locations close to roads doesn't match your idea of activity in the great outdoors, perhaps this is something for you. Confusingly, one category allows operation from a vehicle, but here we're concentrating on the true backpackers' category in which the maximum power allowed is 5W, and there are also limitations on the size of the antenna.

The rules require that you must be at least 100m from your vehicle. However, since the rules effectively limit the size and weight of the equipment, many operators choose to operate their stations much further from a road. Just walking a kilometre or so, even though the route will probably be uphill, could allow you to operate from a much more favourable location. And if you're up for a hike to some lofty peak, you could have a lot more success than your competitors who stick to the lowlands.

There's a constructional elements to this too, albeit primarily mechanical construction. With off-the-shelf antennas you might have to make a compromise between portability and performance. However, all this could change with a bit of ingenuity and a session or two in the workshop. Indeed a high-gain antenna that's lightweight, can be broken down for easy transportation in a backpack, and will withstand strong winds, might just be possible after all.

3: Outdoor Pursuits

Thing 31 – Join a DXpedition

First of all, a word of explanation. A DXpedition is an expedition, often to some far-flung corner of the globe (hence the "DX", which is amateur radio speak for long distance) to set up and operate an amateur radio station. Commonly, these expeditions involve travelling to a country or other location that's rarely encountered on the amateur bands. So, it provides a service to other radio amateurs who might have been eagerly awaiting a contact with that country or island. For expedition members, though, it provides that rare opportunity to be the hunted rather than the hunter. So, while many radio amateurs often remain unheard by rare DX stations, drowned out by stronger signals, for once they can call the shots. A DXpedition need not involve braving the great outdoors like some of our other things in this chapter do. However, even if a building is available for you to locate your station in, there is certainly an element of travel, and not usually to a tourism hotspot. On the other hand, DXpeditions certainly can, and sometimes do, involve conquering the great outdoors, a somewhat extreme example appearing as Thing 32.

Getting Involved

It's unlikely that you're going to be organising your own DXpedition for some time, especially if your ambitions are for a high-profile affair. Instead, many members of the more prestigious DXpeditions gain their places by responding to an invitation for applications, posted online, or even by having been invited. The requirements can be quite onerous, though, so it's unlikely you'd qualify if you're new to amateur radio. Everyone has to start somewhere, though, so how do you get

Photo 31.1. Billy McFarland, GM6DX and 3 x 20m band, ¼ wave raised feed-point verticals, in a phased configuration, DXpeditioning on the Isle of Man.

a foothold into DXpeditions?

One option for a low-key introduction to Dxpeditions is to tag it on to a holiday. The advantage is that it doesn't require a lot of organisation, but there are a few issues you need to bear in mind. First of all, you need to make sure your licence is valid if you're travelling abroad. For several countries such permission is provided without having to apply,

while in several others you can apply for a reciprocal licence – do check. Also, if you're travelling by air, while it shouldn't be a problem to take your equipment, it would be wise to double check with the airline. And it's probably best to use a small rig that you can take as carry-on luggage. Other family members might not share your enthusiasm for operating on holiday, but so long as you don't overdo it, it could be a rewarding experience. And this doesn't apply only to long haul holidays. We note, for example, that while not ultra-rare, several European countries aren't exactly commonplace as DXCC entities (see Thing 14). For example, Lichtenstein is 225th rarest, Gibraltar is 218th (although licensing is tricky), Andorra is 217th, and San Marino is 216th.

Another option is that your local amateur radio club might organise DXpeditions on occasions. Alternatively, you might be able to join another club's DXpedition, especially if you're happy to offer your services for transportation, cooking or similar. Commonly in the UK, the focus of such DXpeditions will be a rare island off the British Isles – which could be of interest to IOTA or WAB hunters (see Thing 16) – rather than a rare country. Even so, taking part in this type of DXpedition could be an interesting and rewarding experience. First, it could take you to some remote and little-visited part of the UK that you'd probably never otherwise visit. And second, it might be your first step on the ladder towards a DXpedition to a tiny and little-heard-of island in the Pacific Ocean.

Figure 31.2. Stevie Chisholm, GM2TT at the front and Jonathan Bowes, MM0OKG at the rear, operating alfresco on a DXpedition.

3: Outdoor Pursuits

Thing 32 – Transmit from a Rock in the North Atlantic

Right at the start we have to admit that most people reading this book won't feel in the slightest inclined to take up the challenge of this "thing". But we really thought we should include it to show the lengths that some radio amateurs go to, in order to pursue their passion to the limit and conquer the great outdoors. Even if you're not inspired to pick up this particular gauntlet, we rather hope that you might exercise your grey cells to come up with the next big challenge. After all, the amateur community has surely left some exciting possibilities still to be conquered.

Facts and Figures

The rock in the North Atlantic referred to in the title goes by the name of Rockall. Rockall is exactly as we described it – it's a small rock in the Atlantic. It's about 25m x 31m at its base, smaller at the top, and there's barely any flat surface on the island, just a sloping ledge near the summit. It rises 17.15m from the ocean. The nearest permanently inhabited place is North Uist, a Scottish island in the Outer Hebrides 370km away. It isn't exactly a regular jaunt for radio amateurs, so it's justly sought after as a rare island for the IOTA award scheme. In fact, it's reported that, in 1971, MP William Ross, speaking in the House of Commons, claimed that "more people have landed on the moon than have landed on Rockall". We have to acknowledge, though, that this is probably no longer true thanks, in no small part, to the DXpedition described here and a couple of previous amateur expeditions.

2023 Rockall DXpedition

In 2023, an expedition to Rockall became the third DXpedition to the outcrop, and it broke all the records. Led by Cam Cameron, the

Photo 32.1. Nobby Styles, G0VJG on the Rockall Dxpedition in 2023

team also included radio amateurs Nobby Styles, G0CJG and mountaineer Emil Bergmann, DL8JJ. The station was operated over 54 hours, during which time it made no fewer than 7,227 contacts. Going beyond the amateur radio operation, Cam aimed to remain on the rock for a total of 60 days, to raised funds for Royal Navy and Royal Marines charity ABF. Sadly, severe weather brought that attempt to an end after 32 days when he was rescued by Coastguards.

The narrative of the journey to Rockall, and the set up, gives just a hint at the challenge of a Rockall DXpedition and the obstacles that had to be overcome. Transport to the island was via the sailing yacht Taeping and its crew. Even that should, by no means, be underestimated, due to the unpredictable and extremely challenging sea conditions. Needless to say, there was no safe landing place and the Taeping had to be anchored several hundred metres away. The team took a small inflatable boat to the base of the rock, from where Emil swam to a suitable place, getting a foothold on the rock on his third attempt. From here he climbed to the summit, using a full range of climbing equipment, installing ropes for the other team members to make their ascent.

Photo 32.2. The Landpod installed onto Rockall.

Installing the equipment wasn't straightforward either, since the initial plans to transfer radio equipment, food and camping gear from the Taeping using a rope and pulley system provided impractical. Instead, numerous trips were required using the small boat and then hauling gear to the summit on ropes, an exercise that took no less than eight hours.

The previously mentioned facts and figures surely give testimony to the success of the DXpedition. But we'll finish by quoting Emil who described this endeavour as "the Everest of DXpeditions," a description that's surely no exaggeration.

Photo 32.3. The Rockall 2023 Team: Emil, DL8JJ (left), Nobby, G0VJG (centre) and Cam (right).

3: Outdoor Pursuits

Thing 33 – Operate from Mountain Tops

We encountered portable backpack operation in Thing 30 when we talked about the RSGB 144MHz Backpackers' Contest. Although some participants in these contests take the opportunity to benefit from a mountain-top location, others don't. Our subject here is quite different in that, by definition, it involves operating from summits.

Summits on the Air

Summits on the Air (SOTA) is an amateur radio awards scheme that was designed to promote portable operation and, in particular, operation from mountain summits and hilltops. It started in the UK, but there are now associations in many countries worldwide. SOTA provides a list of all recognised summits, running to no fewer than 173,596 entries worldwide. You can see the full list at https://www.sotadata.org.uk, but we can't help but feel there's plenty to keep the keen hiker and amateur radio operator happy for quite some time. In most areas of the world, summits must have an altitude of at least 150m. In addition, a summit must be separated from other nearby summits by at least 150m so that ascending every peak involves a distinct climb. Awards are available both for chasers, that is people who make contact with stations on summits and, our main theme here, activators, these being amateurs who set up a station on a summit. To claim a summit an activator must carry all their gear to the summit, and make contact with at least four other stations, none of which can be present on the same summit. While awards are available for various scores, the ultimate accolade is the Shack Sloth trophy for

Photo 33.1. Ascending Sgùrr Aisdair, the Isle of Skye's highest peak, via this steep slope can give you a good work out, even if you're not encumbered by radio gear.

chasers, and the Mountain Goat trophy for activators.

Unusually for amateur radio awards, activator points are not achieved on the basis of the number of contacts made, often with a multiplier, for example the number of countries contacted. Instead, points are awarded for the number of summits activated, but it's not quite that simple. Summits are graded according to their altitude, with different numbers of points awarded for each altitude band. The altitudes and the scores differ according to the country, but scores can vary from one to ten points. Commonly, there's also a bonus point for winter operation.

A British Perspective

To bring this closer to home for British readers, here are a few more facts and figures about SOTA in the British Isles. We're using the phrase 'British Isles' to mean the United Kingdom plus the Crown Dependencies. In terms of figures, there are 1,219 summits in Scotland, 174 in England, 159 in Wales, 65 in Northern Ireland, five on the Isle of Man, two in Guernsey, and one in Jersey

Interestingly, several of the SOTA peaks in the UK aren't ones that will be household names to those with a keen interest in hiking or mountaineering. The list for England, for example, contains old favourites like Helvellyn in the Lake

Photo 33.2. Ben Lloyd, GW4BML and partner Martha Watkins, MW3MBL, on Beinn a'Charnain, taken shortly after Ben's successful marriage proposal, which occurred on a SOTA expedition. Photo Ben and Martha.

District, Pen-y-Ghent in the Yorkshire Dales, and Kinder Scout in the Peak District. However, while St Boniface Down might not be well known to hikers, this 242m hill on the Isle of Wight would earn an activator a point. Unlike most of the well-known English hills and mountains, you can drive to St Boniface Down, and this raises an interesting point. The rules state that you can't operate from a motor vehicle, and operation in "the near vicinity of activator's vehicles" isn't permitted. Not that you'd be tempted to do that, of course.

Since its launch in 2002, over 1,500 UK radio amateurs have conducted SOTA activations, some of them many times. Over 1,800 UK radio amateurs have taken part as chasers, with some having contacted over 50,000 SOTA activations so far. Scotland still has a small number of never-activated summits, six as of February 2024, including some that would be extremely technically and logistically difficult like the sea stacks in the St Kilda islands. Every other UK summit has been activated at least once, and some many times. The Cloud in Cheshire, for instance, has been activated over 2,300 times.

Photo 33.3. Paul Athersmith, M0PLA on Long Mynd-Pole Bank in the Shropshire Hills, on his Mountain Goat qualifying activation. Photo: Paul.

Photo 33.4. Father & son Tom Read, M1EYP and Jimmy Read, M0HGY, on the summit of El Teide, Tenerife, 3715m ASL. Photo: Tom and Jimmy.

Thing 34 – Enjoy High Tech Orienteering

Let's start with a bit of terminology. Our subject here is ARDF (amateur radio direction finding), otherwise known as radio orienteering, or sometimes fox hunting. However, no foxes or hounds are involved, and a high tech version of kids' games involved in finding hidden objects is, perhaps, a better description. Since that's not a particularly catchy name, though, we'll stick with ARDF.

What is ARDF?

ARDF is a competitive event played out in an outdoor environment, commonly in woodland or forest, for reasons that will soon become obvious. Hidden in that environment are one or more beacons, five in the case of international events. These beacon transmitters operate on the 80m (3.5MHz) or 2m (144MHz) amateur bands. Competitors are given a topographic map of the area, showing the start and finish points but, critically, not the locations of the hidden transmitters. The aim of competitors is get from the start to the finish in the minimum time, having found all, or a specified number of beacons en route. Beacons identify themselves using Morse code and transmit periodically in a pre-defined sequence.

The distance from the start to the finish point might be as little as a kilometre, as the crow flies. This isn't a reflection of how much terrain competitors have to traverse, though, because of having to find the beacons. This is especially true since rules for international events often dictate that no transmitter may be within 750m of the start, nor within 400m of the finish or any other transmitter on the course. Typically, courses are designed for 6-10km of total travel distance. Winning times at World Championship events are often less than 90 minutes for 2m courses, and can be under 60 minutes for 80m courses. It might seem surprising that completing a course is easier on the 80m band than on the 2m band. This is because, although 80m signals are largely unaffected by the local topography, 2m signals can easily be reflected, thereby raising the possibility of misleading directional readings.

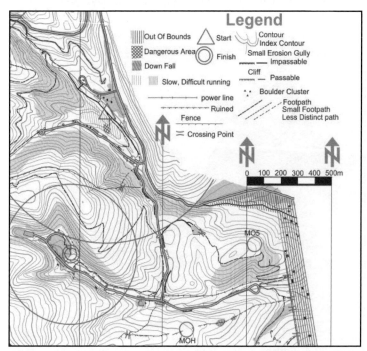

Figure 34.1. Competitors are provided with a map showing the start and finish points. Image: Kharker.

3: Outdoor Pursuits

Technical Skills

Needless to say, ARDF offers an athletic challenge and requires the navigation skills that an orienteering enthusiast relies on. But what about the unique technical challenges of ARDF? First of all, you don't need an amateur radio licence because, although the organisers are responsible for the transmitters, competitors only use a receiver. So, if you're just starting out in amateur radio and have yet to get a licence, this could be for you.

A competitor's receiving station needs a receiver plus a handheld directional antenna and an attenuator. The purpose of the latter is to reduce the received signal strength as you're approaching a transmitter, so that differences in the signal strength, upon rotating the antenna, are still noticeable. And beyond that, you need to hone your skills to use that equipment effectively. In a nutshell, though, competitors usually take bearings on the beacons from a couple of locations to get a first estimate of positions by triangulation. This then allows a route to be planned from start to finish via each of the transmitters. As each beacon is approached, the attenuator would be adjusted to a suitable setting before carrying out another triangulation exercise to better locate that beacon. But it's not just the technical skills needed during the event that are important – success also involves preparation. That preparation involves building up a receiving station that's small and lightweight, easily portable, easy to operate on the move, and rugged. Don't let this deter you if you're just starting out, though. Many first timers gain their first experience of this most unusual form of amateur radio operation using off-the-shelf equipment such as handheld VHF transceivers or scanners.

Photo 34.1. ARDF requires athletic skills, navigation expertise, and skills with operating the radio equipment. Photo: Dale Hunt WB6BYU.

Photo 34.2. Beacons like this on the 2m band are the targets of your search. Photo: Dale Hunt WB6BYU.

Thing 35 – Serve the Community

Amateur radio has long had a tradition for serving the community in times of need. In fact, the amateur radio licence makes specific provision for third-party communication on behalf of the emergency services. And while we might hope that the requirement for such a service is rare, groups of amateurs around the word aim to be prepared, should the need arise.

Natural Disaster Response

A couple of recent high-profile incidents give some idea of the benefit of portable emergency communication when disaster strikes.

Following unprecedented rainfall, widespread and catastrophic flooding occurred to parts of Germany, Belgium, and the Netherlands in July 2021. In Belgium, matters were made worse because the Police building in Wavre, including the computers and antenna of their Tetra communication system, were left completely underwater. What's more, this system is used by fire fighters, doctors, police, civil protection, the army, and crisis coordinators. There were also several local electrical blackouts in Liège, Luxembourg and Brabant Wallon. The 112 emergency number despatch team requested the help of radio amateurs, and about 30 volunteers were deployed in the provinces of Brabant Wallon and Hainaut. They connected fire stations, ambulance stations, hospitals, medical emergency vehicles, the main command post in Wavre, and 112 dispatch in Mons. According to Marc Lerchs, Information Director, Walloon Brabant Crisis Centre, speaking to Crisis Response Journal, "Everything worked perfectly, in VHF, without relays, internet or relying on the official power network, as ham radio operators are self-sufficient and their equipment is battery operated."

In February 2023, a magnitude 7.8 earthquake hit parts of Turkey and Syria. Tragically, huge numbers of people lost their lives, and whole areas were totally destroyed. What's more, failure of the usual communication systems had hampered search and rescue efforts. It's encouraging to note, therefore, that Turkey's amateur emergency communications group Türkiye Radyo Amatörleri Cemiyeti (TRAC) coordinated primary communications. Furthermore, we can report that search and rescue groups were advised to have at least one amateur radio operator in their team and/or in their facilities. We must assume that many people owe their lives to this endeavour.

Photo 35.1. Radio amateurs in Belgium made an important contribution to rescue efforts during flooding in 2021. Photo: Christophe Licoppe, European Commission.

3: Outdoor Pursuits

Introducing RAYNET-UK

According to its website, "RAYNET-UK is the UK's national voluntary communications service provided for the community by licensed radio amateurs". Furthermore, "RAYNET has provided additional communications at major incidents involving aircraft, trains, flooding, evacuations, telephone exchange failures, missing persons searches, adverse weather, and oil/chemical pollution". Thankfully, emergencies in remote areas, or where the incident has damaged the communication infrastructure used by the emergency response services, are rare in the UK. Nevertheless, RAYNET-UK remains prepared and, quite appropriately, this emergency preparedness is maintained through a quite different form of community service. By providing communication services for organisers of a range of outdoor events, such events also act as training exercises for group members.

Photo 36.2. RAYNET taking part in a simulated crash on the North Yorkshire Moors Railway, an exercise also including Police, Fire, Ambulance, Mountain Rescue, Yorkshire Air Ambulance, RAF and Environment Agency. Photo: Jon Woollons, G1OSP.

Formed in 1953, RAYNET-UK has grown into an organisation with around 5,000 members, providing communication assistance on many hundreds of events each year. Included here are car rallies, canoe races, mountain biking events, marathons, triathlons and sponsored walks. A classic example is the annual Yorkshire Three Peaks Race, the route of which takes in the summits of Pen-y-ghent, Whernside and Ingleborough. Covering a distance of 37.4km, the race involves 1,609 metres of ascent and descent. Following a tragedy in severe weather conditions in 1978, in which runner Ted Pepper died, revised safety regulations and race control procedures were introduced. And so we come to RAYNET-UK. All communications between Race Control in Horton-in-Ribblesdale and checkpoints, including the summits of Pen-y-ghent, Whernside and Ingleborough, are provided by RAYNET operators. This provides control on the progress of competitors and safety information from remote locations to alert officials to any potential problems.

Photo 35.3. RAYNET provides support during the Three Peaks Race on the summit of Peny-y-ghent in the Yorkshire Dales. Photo: John Jebb.

Thing 36 – Communicate with Cavers

Caves and other underground places are surely some of the most challenging of environments for radio communication. Loss of mobile phone coverage on entering a road or rail tunnel are testimony to this, and the GPS signals required for navigation are also blocked. Despite this, radio amateurs have turned their attention to this most unforgiving environment, and not just as a technical challenge. In some areas, radio amateurs do indeed aim to overcome obstacles just because they're interesting challenges, even if there's no apparent real world application, but not here. Instead, the main motivation is to serve the community and, in particular, the caving community. This might be the last of our "things" concerned with the great outdoors but last doesn't mean least. In fact it's longer a read than the other outdoor-related topics because it involves more of a technical challenge. Also, this prominence might just have more than a little to do with the fact that this subject is a passion of your scribe.

The fact that caves are potentially dangerous doesn't go unnoticed, thanks to the not infrequent news reports of caving incidents. When cavers fall, get lost, or are cut off by floods, the members of a volunteer rescue team are called out with the aim of bringing them to safety on the surface. In the case of serious injury, the availability of medical assistance at the cave entrance can make the difference between life and death. Every second counts, and extricating a casualty on a stretcher along narrow twisting passages and up vertical shafts can be a slow process. So being able to summon a medical presence before the casualty reaches the surface is essential. Meeting this communication requirement is one of the

Photo 36.1. Volunteer cave rescue teams are the main beneficiaries of the development of cave radio equipment and techniques. Photo: Bartek Biela.

main reasons that radio amateurs have developed equipment and techniques that can communicate between a cave and the surface.

Through-the-Earth Radio

We might have rather suggested that radio can't propagate through rock, but that's a gross over-simplification. In fact, attenuation through conductive media decreases with decreasing frequency. And in the case of the limestone, in which most caves are formed, a frequency around 100kHz is effective. Indeed, cave rescuers have used radios on a frequency of 87kHz since the 1970s. Because of the long wavelength at these frequencies, efficient antennas aren't feasible, so small loop antennas were used initially. They didn't actually radiate a signal, in the strict sense of the word but, instead, they generated an induction field. The upshot of this is that the signal decayed with the cube of distance instead of the inverse square law that applies to true radio. As a result, the range was limited to around a 100m or so. Even so, this was adequate to cover the distance between an underground location and the surface in many cases, but not always. So, when these radios started to fail, and the rescue teams were looking for a replacement, radio amateurs came to the rescue.

Amateur experimentation had showed that a much greater range could be achieved using a ground antenna. In other words, the 'antenna' comprised a pair of electrodes separated by a few tens of metres. This also had the benefit that, since it

Photo 36.2. Radio amateurs, like Fred Rattray, G4SPR here in Skirwith Cave, have played an important role in pioneering cave communication. Photo: Chris Hunter.

Photo 36.3. Here we see the HeyPhone, designed for rescue use by John Hey, G3TDZ, in its rugged and waterproof Pelicase.

operates mostly via ground conduction, Ofcom didn't consider that its use should be regulated. To cut a long story short, this form of cave radio became a practical proposition in the form of the HeyPhone. Designed by the late John Hey, G3TDZ, this cave radio served the rescue community well from 2001 until quite recently.

In passing, the comparatively short time that cave radio designs remain in operation is due to the unforgiving nature of the cave environment. Despite efforts to make equipment rugged and waterproof, knocks, drops and dunkings eventually take their toll. Such is the challenge of cave communication.

More recent designs have now replaced the HeyPhone, and are serving the rescue teams well, but radio amateurs have worked on alternative solutions in the meantime. Even with the use of a ground antenna, in many cases the limited range means that the surface station has to be located immediately above the underground station. Needless to say, though, being able to remain in contact from a vehicle on the road is much more preferable to working from a rainy and cold location on a wind-swept hillside at night. With this in mind, radio amateurs investigated the use of HF radio, despite conventional wisdom suggesting that attenuation would be too high. Without getting into the technical details, just let's say that at these frequencies the rock is no longer a 'good conductor' so the relentless increase of attenuation with frequency starts to tail off. And if HF radio could be made to work, it was argued, it would be able to generate a true radiated signal and this would give rescue teams more freedom in locating the surface station. Not only that, but it might also open up the possibility of being able to transmit warnings

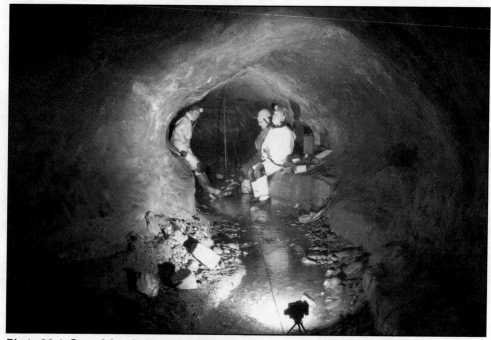

Photo 36.4. One of the challenges of HF cave radio is erecting an antenna and making sure other cavers don't trip over it. Photo: Gregory Collins.

3: Outdoor Pursuits

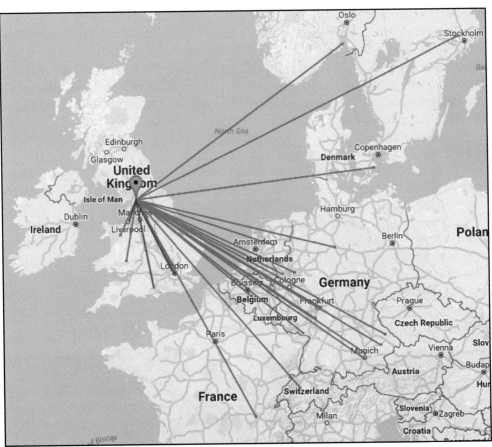

Figure 36.1. The potential of HF as a vehicle for proving warnings to cavers was demonstrated by receiving WSPR signals underground from across Europe. (Map Data ©2016 Geo-Basis/DE-BKG (©2009), Google Inst. Geogr. Nacional).

to cavers – for example, 'heavy rain on the way' – over a large geographical area.

Underground HF antennas are problematic, but these problems are not insurmountable. Antennas for the 40m (7MHz) band are just about manageable, thanks in part to the fact that their length is significantly reduced due to the effect of the surrounding rock. Mostly, inverted-V antennas have been used, although they sometimes have to be far from straight due to the meandering nature of cave passages – see **Photo 36.4**. Initial tests were encouraging, with communication from a location 100m underground being maintained with a surface station up to 5km away. More recently, WSPR has been used to better characterise propagation. Indeed, from that same underground location, signals were monitored from stations around Europe, the most distant being in Sweden at a range of 1,327km (see **Figure 36.1**). This gave some credence to the concept of a wide area warning system. And while WSPR signals transmitted from this cave weren't logged from any surface stations, a modicum of success was achieved from a different, shallower cave. Indeed, signals from a Yorkshire Dales cave were received in Liverpool, London and Southampton, the latter being 370km away. Some success was also achieved in transmitting from Yordas Cave, also in the Yorkshire Dales, even though this was largely for fun, as opposed to because it proved a genuinely valuable technique. The cave has a huge chamber which is close to the entrance and at a shallow depth, and this permitted a full-sized 80m inverted-V antenna to be installed. The bottom line is that an SSB contact was made with a station in Ipswich, at a range of 340km.

More recent work by radio amateurs has involved using some of the digital modes of communication that have reaped benefits in the HF bands (see Thing 10). So far, good results have been achieved using PSK31 – although using a microphone is easier than typing text with fingers numbed by cold water – and SSTV signals have been transmitted from underground to the surface.

As a final comment on through-the-earth radio, it's appropriate to mention a quite different application, which should strike a chord with those interested in amateur radio direction finding (Thing 34). Although it doesn't work with ground antennas, an underground radio transmitting into a loop antenna can be found on the surface by triangulation. And having found ground-zero, the depth can be determined by measuring the angle of the field lines which are like those surrounding a bar magnet. This type of radio-location can be used to correct some of the errors

Photo 36.5. This location isn't a typical underground environment, but the chamber could house a full-sized inverted-V antenna which allowed an SSB contact over 340km. Photo: Ian Cooper.

3: Outdoor Pursuits

that accumulate when surveying a cave using conventional instruments.

Along Passage Radio
Through-the-earth radio isn't the only option for cave communication, and radio amateurs have conducted tests into several alternatives. One of these is referred to as guidewire communication, and can be thought of as a simpler and much cheaper version of the leaky feeder systems that are often used in transport tunnels. It involves installing a wire along a passage from the surface to the underground location, and communicating using handheld radios held close to the wire. In one sense it doesn't altogether sound like true radio, since it relies on a wire to propagate the signal, although it has been used successfully. But while some rescue teams continue to use communication systems that involve wires, these method aren't ideal because of the work involved in laying the line before communication is achievable.

Radio can also propagate along cave passages even without a guidewire, and

Figure 36.6. Underground SSTV has been demonstrated by transmitting images to the surface from a historical lead mine in Shropshire.

Photo 36.7. Rob Gill, G8DSU, triangulates an underground beacon as part of the radio-location exercise.

101

that has also been investigated. Propagation is via a lossy waveguide model, which means that the frequency has to be above the so-called cut-off frequency. In the case of most cave passages, the implication of this is that microwave frequencies have to be used. Even so, while a range of, perhaps, a few hundred metres could be achieved in some cave passages, the range is often much less due to the passage's small cross-sectional diameter, wall imperfections such as roughness and, most significantly, bends. A single bend can attenuate the signal to such an extent that two bends in the signal path is often a show-stopper. Nevertheless, multiple repeaters can be used to extend the range and, with this in mind, work has been carried out into modelling propagation at a range of frequencies in a variety of cave passages, as a tool to assist the design of such a multi-repeater system.

Photo 36.8. Mike Bedford, G4AEE, characterising attenuation vs distance on the 23cm band in Kingsdale Mater Cave in the Yorkshire Dales.

4

A Personal Perspective

Introduction

A strange title for a chapter, you might think, so let's start by explaining what we mean by a personal perspective. In fact, there are two strands here, both of which are integral to amateur radio, and both of which have a personal element. All our "things" here relate to one or both of these strands.

First of all, to counter any view that amateur radio is purely an individual activity, we'll look at several ways in which amateur radio is, or can be, a social activity. Indeed, there are aspects of amateur radio that are primarily group oriented. We've seen some of these already, mostly in our discussion of outdoor pursuits, and here we offer some additional "things" that are either fundamentally a group activity, or have an important social or community element.

Next up, we use the word 'personal' to refer to "things" in amateur radio that can help you become a more rounded personality by developing new interests, or even to help you in your career progression.

The chapters on operation and construction might seem like the bread and butter of amateur radio, and so they are, and the chapter on outdoor pursuits has a fairly obvious theme. This chapter might appear altogether different in that it is not quite as obviously related to a technical hobby. So, rather than prevaricate any further, we'll jump straight into our next few "things" as we delve further into the astonishing diversity of amateur radio.

Thing 37 – Meet Like-minded People

Amateur radio is unusual among technically-oriented pastimes. In particular, it's surely rare that a technology-based hobby has a social element at its very core. Of course, this doesn't apply to all aspects of amateur radio, in fact few if any of the "things" we looked at in Chapter 2, which concerned construction, normally lend themselves to cooperation. For example, we struggle to think of how soldering components onto a circuit board could be considered a social activity. We have already seen some group activities, though, and here we could mention operating in a field day (Thing 30), taking part in a DXpedition (Thing 31), competing in amateur radio direction finding events (Thing 34) and serving the community (Thing 35). However, we'd like to suggest that there's also a social element at the very heart of amateur radio. After all, this many-faceted hobby is all about communication, and specifically communicating with people in the form of other radio amateurs. For some people, the communication element is all concerned with testing out their latest antenna or seeing how far they can transmit with a new super-simple, low-power transmitter. And there's undeniably an element of this to most amateur radio operation. However, if you also relish the opportunity to engage with the people behind the callsigns, we offer some guidance on how to do that.

Spotlight on Modes

In referring to modes we're talking about the methods by which information is transmitted and received, and we looked at several in Chapter 1, as part of our discussion of operation. Included here are obvious modes such as speech, Morse code (Thing 10), and the various modes for exchanging textual information (Thing

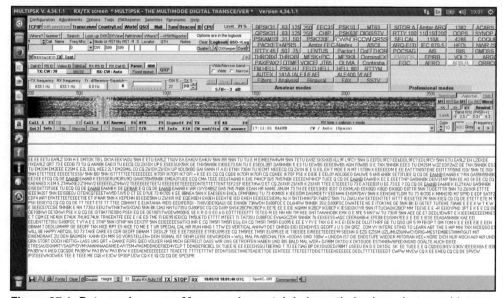

Figure 37.1. Data modes, even Morse code, certainly have their place, but vocal intonations are sadly missing.

11). And to take an extreme case, modes such as JT65 that we discussed in Thing 12 don't provide much scope for getting to know someone. They have their place, of that there's no doubt, in allowing communication that just wouldn't be possible with other modes. However, if you can only send 13 characters, and even that takes 50 seconds, you're not going to be discussing the weather or your pet cat.

Other textual modes make more normal communication possible in allowing free-form conversational exchanges. In this respect they're similar to mainstream methods of textual communication such as email, text messaging or WhatsApp, but without the option of adding images, movies, web links and the like. While this is a big improvement over JT65 and other terse fixed-format modes, it's hard to deny that the ultimate in inter-personal communication can only be found via face-to-face communication. After all, facial expressions and vocal intonations are missing from all other methods of communication, yet they add so much to the exchange. Amateur TV (Thing 43) is about the only mode that offers all of this, but it's technically challenging and is mostly limited to the UHF bands and above, so ranges are limited. The second best is surely voice communication, and here the world's your oyster, surely a big advantage if you want to foster international links.

Photo 37.1. If you really want to engage in meaningful conversations, speech communication surely takes some beating. Photo: Rrbhosale.

So, by all means use JT65, RTTY or Morse code. All of these have their place. But if you really want to get to know people, verbal communication surely takes some beating.

Practical Guidance

So much for the technical aspects of communication modes, let's now look at some practical aspects to help you get the most from interaction with radio amateurs around the globe.

First up, you need to be sensitive to other people's preferences and aspira-

Photo 37.2. Photo: If you're looking for long ragchews, tuning your radio to the 3.5MHz band during the day will offer plenty of opportunities.

tions. Some people want to make as many contacts as possible, and this means restricting the exchange of information to little more than callsigns and a signal report. Indeed, this is to be expected during a contest, as we saw in Thing 13. In the main, though, outside of contests people will exchange their name and location, and perhaps some information about their equipment, but rarely much more. However, if you hanker after a bit more of a normal conversation –and it has to be admitted that it would be rare to meet someone at an in-person social event and engage in such a short conversation – listen around before you call someone. This way, you'll get more of a feel for who sounds like they'd be up for a more protracted exchange.

Something else to bear in mind is that some of the amateur bands, particularly at some times of day, will offer you more opportunities than others for getting to know people. By and large, these will be the lower frequency bands such as 160m (1.8MHz) and 80m (3.5GHz), especially during the day, when they will offer reliable communication over, say, 150km to 800km. Because the opportunities for long-distance contacts are almost non-existent in daylight hours, those who are on the band will probably be there to engage in more protracted contacts, or 'rag-chews' as they are often referred to. What's more, in many countries, this limited range will mean that you'll be contacting people who speak the same language, which will also improve the ease with which you'll be able to have a more meaningful conversation.

This reference to the language barrier is an interesting one and, of course, it shouldn't be taken to mean that longer chats with foreign amateurs aren't possible. The proliferation of the English language, and especially among radio amateurs, is a big advantage to those of us who are native English speakers. And there are surely amateurs who use their hobby to improve their skills in speaking English. But if you consider yourself a linguist, don't discount amateur radio contacts as a means of brushing up on your skills. In fact, while not having tried it myself, so this is purely speculation, it seems reasonable to believe that some non-native English speakers would welcome the opportunity to chat with you in their own language.

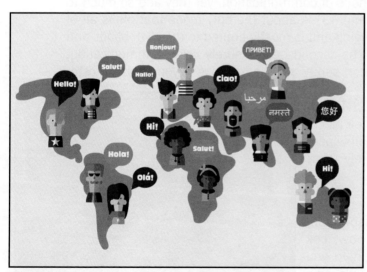

Figure 37.2. Language differences needn't be a barrier to effective communication and could offer you the opportunity to hone your linguistic skills. Image: 1940162 Hari chandana C.

4: A Personal Perspective

Thing 38 – Join a Local Club

If having meaningful conversations with like-minded people via the amateur bands is good, you might consider that meeting up with fellow radio amateurs in person is even better, and we'd tend to agree. If this is your view, we recommend that you consider joining a local amateur radio society. And while we're presenting this as an opportunity for social contact, you'll probably find that you'll learn a lot from being part of a club too. You'll probably choose to join your closest club, but if you live in a densely-populated area, you might find that you've got several within travelling distance to choose from. So, what might you expect of an amateur radio club? Let's take a look, but do bear in mind that clubs differ, so our overview here almost certainly won't be fully representative of all clubs.

Regular Meetings

Club life is usually centred around regular meetings which take place, typically, every one or two weeks. Commonly, these meetings involve a talk, perhaps illustrated with a demonstration. Topics could be instructional, perhaps on a technical aspect of amateur radio, or they might be inspirational, for example a presentation of someone's experience in a special interest within the hobby. Talks may also be on a general interest topic on a subject that's peripheral to amateur radio. For example, I've attended a talk on the emergency services' communication systems, and I've given several talks on cave communication (Thing 36), not all of which involve amateur radio in the strictest sense of the phrase.

It's not all talks, though. Some clubs run construction projects for members. While construction is normally an activity carried out in isolation, there are benefits in a group project. For example, short talks or demonstrations might set the scene for the next phase of the project. Alternatively, those experiencing difficulties can bring their partially-completed projects to a meeting to solicit the assistance of more experienced members.

Some clubs meet in a pub or other third-party venue. A few have their own premises, and this permits them to have a club radio station. Using the station at a regular meeting can provide motivation for those who have yet to obtain a licence. It can also offer operating experience to those who either don't have a home station, or have a station that's limited in

Photo 38.1. Club night in Braintree

some way, perhaps due to lack of space for antennas.

Other regular meeting themes could include quizzes, bring-and-buy sales, Morse code training and practice, and preparation for other special activities as discussed below, to name but a few possibilities. Meetings may also provide a less formal opportunity for members to discuss a whole range of topics, often in a social setting.

Other Activities

Activities outside the regular programme of meetings are many and varied. Clubs commonly arrange scheduled on-air meetings for members, on particular frequencies and at specified times. Some clubs offer training sessions for those intent on gaining an amateur radio licence or perhaps upgrading from one level of licence to the next. This is an opportunity, both for would-be licence holders and for trainers – see Thing 41.

Field day contests (Thing 30) are usually organised and operated by clubs as opposed to individuals because of the challenge of setting up and operating a portable station for 24 hours. The scope for collaboration – both in the preparation and over the weekend of the event – is not hard to appreciate.

Another activity commonly undertaken by clubs is to arrange a special event station to demonstrate amateur radio to the general public. As you can read in Thing 40, special event stations celebrate an event which is of special significance, and will be open to viewing by members of the public. Radio clubs will often set up these stations at events that are of special significance to their locality.

Joining a local club is popular with those across the entire spectrum of experience, but club membership is particularly beneficial for those who are just starting out in amateur radio. For a start, talks on amateur radio, seeing the club station being operated, or just chatting to more experienced members provide the inspiration that many would-be amateurs need to take the next step. That next step, of course, is obtaining a licence, And, if you're lucky enough to find that your local club organises training sessions to help you pass the necessary examination, you could soon be on your way to the amateur bands.

Special Interest Clubs

As a parting shot on clubs, we ought to refer to a different type of amateur radio club. Our emphasis so far has been on local clubs which cater mainly for mainstream interests. If you have a particular interest in a more-specialised subject, though, you might find that there's a club that caters for that interest. These clubs will mostly be national societies, as opposed to local clubs, so the opportunities for person-to-person contact is more limited, so this subject warrants only a passing mention in this chapter. Suffice to say, though, you'll find special interest groups for Amateur TV (Thing 43), QRP (low-power operation), Microwaves (Thing 6) and Summits on the Air (Thing 33), to name but a few.

Thing 39 – Discover the World

First published by RSGB in 1967, World at Their Fingertips: the Story of Amateur Radio... and, A History of the Radio Society of Great Britain by John Clarricoats, G6CL, was considered an influential book, introducing many to the hobby. And the sentiment is just as true today – radio amateurs really do have the world at their fingertips. What's more, making contact with amateurs around the globe can certainly provide a new appreciation of our planet. Around the time I first started to get interested in amateur radio in my mid teens, I had little interest at all geography. How that changed when I started listening on the amateur bands and especially when I obtained a licence. I still remember discovering that Rhodesia – now Zimbabwe – wasn't a European country after all. So, do you want to discover the world? Amateur radio could well provide a new appreciation of Planet Earth and perhaps even inspire a wanderlust in you. You might even end up combining your new-found interest in world geography with amateur radio by taking part in a DXpedition, as we saw in Thing 31. And even if you don't aspire to those dizzy heights, as we discussed in the same thing, you could try operating from overseas as part of a holiday.

A Voyage of Discovery

If the idea of discovering the world, courtesy of amateur radio, is an appealing one, how should you go about it? In one sense, you don't need to do anything, it'll just happen as you're exposed to amateurs around the word. However, if you want to be rather more proactive, here are some practical tips.

Just talking to someone in Jakarta could well reveal that it's the capital of Indonesia. However, unless you discover someone unusually chatty, it might not help you to appreciate that the country is an archipelago of over 17,000 islands between the Indian and Pacific Oceans. Nor would you learn which of all those islands Jakarta is located on – actually it's on Java. The key here, of course, is to use a map and most logging software allows you to export data in a format that allows you to import it into your favourite mapping software. However, rather than plotting contacts on a map as an afterthought, it's probably better to see

Photo 39.1. There's only so much of our astonishing world that we'll see in person, but we can learn a lot about it through amateur radio.

Photo 39.2. It might seem 'old school', but how about hanging a paper map on the wall and marking some of your contacts with pushpins. Photo: Mark Levin.

a world view in real-time. And some logging software does, indeed, allow you to see the location of the station you're currently in contact with on a world map. However, there's still something rather special about good old-fashioned paper maps. So, how about putting a world map on the wall of your shack – that's the room housing your station – and using pushpins to highlight some of your most interesting contacts?

We recognise that most amateur radio contacts involve the exchange of minimal information. But, as we discussed in Thing 37, if you hanker to discover the people behind the callsigns, do look for opportunities to engage in more meaningful conversations. And those conversations could well allow you to learn more about those far-away places. As an example, it's more years ago than I'd like to remember, but I recall speaking to someone in Ecuador, who announced "it's a glorious morning here in Quito and the volcanoes are out." I had assumed that he was saying they were erupting but, in retrospect, it seems more likely that he was telling me that Catopaxi, and other nearby volcanic peaks, were clearly visible.

Photo 39.3. Even a casual comment heard on the amateur bands can give more of a feel for a place that a geography textbook would. Photo: Agencia de Noticias ANDES.

4: A Personal Perspective

Thing 40 – Introduce the Public to Radio Technology

Fancy the opportunity to showcase amateur radio to the public? If so, how about setting up and operating a special event station? You could do this individually, or with one or two friends, but it's far more common, and probably more rewarding, to undertake this activity as a club event (Thing 38).

Special Event Stations

A special event station is an amateur radio station that is set up to support a special occasion such as a local or national event or a celebration or anniversary. Members of the public will be invited to visit the station to see amateur radio operation at first hand. In fact, in the UK, both the special occasion and the invitation to members of the public are requirements for obtaining a special callsign, something we look at later. Special event stations operated in the UK during 2023 included ones at special events at museums, at primary schools in support of British Science Week, at classic car shows, scouting events and village fetes, fairs and festivals plus, of course, celebrations of the coronation of King Charles III.

Special event stations are able to use special callsigns. In the UK, these differ from ordinary full callsigns in having the GB prefix followed by a figure and two or three letters. Those letters usually relate to the event being supported. So, for example, an event called Crank It Up! took place at Beamish Museum during 2023, and it offered visitors the opportunity to "Unleash your inner inventor with amazing science, technology, engineering and maths activities." Bishops Auckland Radio

Photo 40.1. Intoducing the public to amateur radio

Photo 40.2. Le Mans, a historic event

Amateurs Club supported this initiative by hosting a special event station with the callsign GB4BM, the "BM" standing for Beamish Museum. While a special callsign will surely go over the heads of most visitors, their use does make the station somewhat of a rarity and, in so doing, will probably increase the number of contacts made by the station. The available callsigns for special event stations differ around the word and, in addition to having a different prefix, might allow more than the usual single figure. For example, a French special event station to commemorate the 100th anniversary of the 24 Hours of Le Mans endurance motor race used the callsign TM100GPF. In fact, here in the UK such bizarre callsigns are permitted for the oddly-named special special event stations, but such callsigns will only be granted for celebrations of major national events.

Practical Advice

Arranging a special event station is easy enough, given sufficient support from club members. Organising one that will engage and fascinate visitors is rather more challenging. Here are a few pointers to making the event more successful in that respect.

First of all, give some thought to the design of the station. In particular, you should aim for more than an operator sitting at a table containing the radio equipment. A selection of display stands would certainly brighten it up. Their design is limited only by your imagination but could show the name of the organising amateur radio club, the station callsign, bullet points about amateur radio, and a good selection of photos illustrating something of the diversity of amateur radio. Don't omit what you think is obvious, so it wouldn't be inappropriate to use a headline on the lines of "Amateur Radio Station GB0ZZZ".

Consideration should also be given to allowing the viewing public to see something more enthralling than the operators' backs. Normally this means they should face the public, although that would mean that all they see of the equipment is the backs. One solution here is to set up a camera and monitor showing those parts of the station that would otherwise be invisible.

Don't assume that visitors will understand what's happening – they probably won't. Instead, make sure you tell them. One method is to avoid operating continually, thereby allowing the operator to take periodic breaks during which they can provide an explanation for your visitors. Alternatively, you could have someone mingling with the visitors explaining what's happening. Above all, though, have fun and hopefully that enthusiasm will be infectious. So, be sure to choose people who genuinely enjoy engaging with the public to take part in these events.

Thing 41 – Help People Gain a Passion for Radio

In the maker community – which we can think of as the hobby of electronics brought into the 21st century – training and encouraging newcomers, especially young people, is integral. So, we find coding clubs to teach programming, and after school clubs covering subjects from robot building to electronics and 3D printing. Central to the maker movement are hackspaces which, according to hackspace.org.uk are "places you can visit to meet people, learn, socialise, and collaborate." The "learn" aspect is important, indeed many hackspaces offer classroom talks and hands-on training. The amateur radio community also has an element of teaching, training and encouragement at its core so this is an opportunity for those who want to enthuse a new generation. And this goes beyond drawing young people into amateur radio. STEM subjects (science, technology, engineering and maths) are vital, both for business and industry, and for giving young people an opportunity to engage in a stimulating and rewarding career. So, if you fancy playing a part in this worthwhile activity, here's how you can do just that through amateur radio.

Photo 41.1. Pupils at Sandringham School being intoduced to amateur radio

Informal Training

You might not feel confident in giving talks or teaching classes, but this doesn't mean you can't play an important role in sharing your knowledge. You certainly shouldn't underestimate the worth of just being around newcomers. After all, enthusiasm is infectious, and knowledge will rub off without you even knowing it.

Of course, this is only possible if you come into contact with the next generation of radio amateurs, and here we're thinking particularly of those who have yet to obtain a licence. Local amateur radio clubs (Thing 38) can play an important role here. We certainly recommend that those newcomers who have a spark of interest in amateur radio make themselves known to local amateurs by attending a club meeting, and ideally joining a club. So, do look out for new faces at your club and do your best to make new people feel welcome.

Whereas newcomers will pick up a lot just by being around you and other club members, if there's a club station that operates on some club nights, this too has an important part to play. After all, just watching and listening, to find out what it's like to operate an amateur radio station, can impart so much information that amateurs take for granted. And don't forget that, so long as a licensed amateur is present to supervise, you can also invite unlicensed people to speak on the amateur bands. What better way could there be to provide them with the motivation to obtain a licence of their own.

Examination Classes

Amateur radio licences are only granted by those who have proven their competence to operate an amateur radio station by taking a passing an examination. There are several ways of gaining the knowledge needed to pass these examinations. Books for self teaching offer one option – the RSGB publishes books for each level of licence examination, namely Foundation, Intermediate and Full – and these are recommended, even for those undertaking more formal training. There are also online courses, many of which are run by local clubs. These can either be in-person via Zoom, or they could be self-paced courses. If your club runs such courses or is considering doing so, this is an area in which you might be able to contribute.

Some clubs also offer in-person courses, and these offer the additional benefit that attendees have the opportunity to learn alongside other would-be amateurs. If your club runs courses, and if you consider yourself a communicator, how about offering your services in running some of the sessions. Alternatively, if your club doesn't currently run a course but you've recently had an influx of new members interested in obtaining a licence, this might offer you a future opportunity for rewarding involvement.

Photo 41.2. Heads down for the exam

4: A Personal Perspective

Thing 42 – Improve your Employment Opportunities

Could an interest in amateur radio improve your employment opportunities? This is a question that has been debated over the years, so it's hardly surprising that there's no simple answer. One thing's for certain, and that's that it's not a universal panacea, and there are many careers in which it will offer you no benefit at all. However, there's another thing that we are equally sure of, and that's that in other areas it can indeed improve your employment opportunities. What's more, this isn't speculation, in fact your scribe can testify to this from personal experience. However, before playing the amateur radio card in applying for a new position, there are several things you need to consider, and that's our subject here.

Job-Specific Considerations

They might be few and far between, but a job in amateur radio might seem like a dream come true to the amateur. It might be stating the obvious, but if you do see a job advertised with a company involved in amateur radio, your hobby will almost certainly help. So, for example, being involved in amateur radio will probably be an essential requirement for a sales position with a retailer of amateur radio equipment. Even so, be sure to take a look at the section on writing a CV for guidance on how to put across the fact.

At the other end of the spectrum are positions in which engaging in amateur radio probably won't give you an edge. And, unfortunately, these jobs are in the majority. Even so, if you're applying for a job in financial management, for example, don't necessarily assume that your amateur radio experience is of no value, as we discus later.

Between these two extremes are technical positions in which involvement in amateur radio could be beneficial. Included here are roles in electronic design, especially if there's an element of analogue design, and even more so if that's RF design. Even so, you need to be careful about how you sell your hobby, and this brings us to our next topic.

Photo 42.1. A technical role is the most obvious type of position for which an interest in amateur radio will be beneficial. Don't write off your hobby in applying for other positions though.

CV Preparation

Whether or not hobbies should be included in CVs is another question that has long been debated. However, the general consensus is that their inclusion can be beneficial, but only if the hobbies have a bearing on the advertised position. Interestingly, though, this doesn't necessarily mean that amateur radio shouldn't be included in a CV for a job with no technical content. However, it has to be handled carefully, especially since the uninformed can have some wrong perceptions of amateur radio.

As a general rule, it wouldn't be beneficial to add a terse bullet pointed statement such as 'Amateur Radio (callsign G4AEE)' under a heading of 'Hobbies and Interests'. Instead, reference to a hobby should relate to the advertised position and, better still, it should refer to a requirement specified in the advertisement. Furthermore, it's often suggested that work experience is much more important than hobbies, so any hobbies should only be listed if they illustrate a skill in which you have no work experience. For example, in applying for a first position on the management ladder, it might give you an edge if you're able to say 'Team building skills obtained as president of an amateur radio society'. The same applies to more technical positions. But, again, rather than just making a brief reference to amateur radio, something like the following should be considered: 'Analogue design experience obtained through an involvement in amateur radio'.

Finally, if you have particular skills that you've honed through amateur radio, they might lead you to an unadvertised position, and I speak here from my own experience. This relates to my hobby interest in sub-surface communication for use by cavers and cave rescue teams as discussed as Thing 36. To cut a long story short, that experience led to me being invited to an interview and subsequently receiving a job offer for a research position with a company concerned with mine safety. In turn, this led to a senior research role in academia. Admittedly, I was probably in the right place at the right time, but hopefully it provides some encouragement if you do have aspirations to turn your hobby into a career.

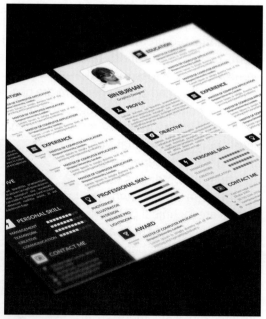

Photo 42.2. With careful handling, reference to amateur radio in your CV could give you an edge.

5

Beyond the Everyday

Introduction

Hopefully the 42 "things" we've seen so far have illustrated something of the diversity on offer from the hobby of amateur radio. We rather expect that some of them will have been eye-openers. In this chapter, though, we go one further in presenting some additional "things" you can do with amateur radio that are, perhaps, even more unexpected.

Most of these "beyond the everyday" topics are concerned with operation, but will utilise some uncommon methods of getting a signal from A to B or involve transmitting less common types of information. As a result, they require some more specialised skills. They probably aren't for beginners, therefore, but even if you are new to amateur radio, we trust that you'll find them interesting. What's more, perhaps they'll motivate you to start out on a path that might eventually lead to these non-everyday "things".

Two of the things are concerned with personal development. As such, they could easily have been discussed in the chapter on Personal Perspective if it wasn't for one thing. That one thing, of course, is that they're beyond the everyday. So, again, you're probably not going to be achieving these things anytime soon. But that doesn't mean that these are "things" you can't aspire to, such is the diversity and challenge of amateur radio.

117

Thing 43 – Exchange TV Signals

Back in Thing 10, as part of our discussion of methods of transmitting data, in the broadest sense of the word, we looked at SSTV, i.e. Slow Scan Television. As you'll recall, SSTV occupies a narrow bandwidth, no greater than that of a speech signal, which allows it to be used on the HF bands, i.e. the shortwave amateur bands. This, in turn, means that it can be exchanged with stations anywhere in the world. The downside of that, though, is that an SSTV picture has a low resolution, and it takes several seconds to transmit the image. In other words, it is used for exchanging still pictures and not moving images.

Our subject here is what most people would think of as 'proper TV' or, in other words, the exchange of moving images. Despite the fact that amateurs have been transmitting and receiving broadcast standard TV since the 1950s, the latest technical developments mean that full high-definition pictures with sound are now within the reach of the amateur community. Although we referred to broadcast standard TV, and because TV is normally associated with entertainment, that isn't the aim of amateur TV (ATV). While there's nothing to stop you from making entertaining or feature-length recordings for transmission, the terms of the amateur licence still apply, so you are not allowed to "broadcast". As with voice or Morse transmissions, you must only transmit the pictures to another amateur or groups of amateurs with whom you've established contact. Typical pictures are of the operator, or the operator's latest project. So, if you fancy yourself as the next Stephen Speilberg or your favourite film director, this isn't for you. However, if you relish the technical challenges of setting up and operating an ATV station, do read on.

Amateur TV Operation

Most amateur TV is now conducted using digital modes, specifically DVB-S and DVB-S2 as used by commercial satellite TV transmissions. These allow the compression of an HD TV signal into less than 500kHz of bandwidth. The advantage of this is that the signal can fit into smaller spaces on the crowded amateur bands. Tests have taken place on the 4m (70MHz) and 2m (144MHz) bands, but most operation is on 70cm (430MHz) and above. Using even more digital compression, it's possible to transmit standard definition ATV signals in the 6m (50MHz) band and low definition pictures in the 10m (28MHz) band.

Simplex ATV contacts on the 70cm (430MHz) and 23cm (1.2GHz) bands typically reach ranges of up to 50km and the use of ATV repeaters can easily double this range. Portable ATV operation from hilltops typically reaches 150km range but, during enhanced propagation, contacts have been made at over 400km. Satellite ATV operation is very popular using the wideband QO-100 transponder on the geostationary Es'hail2 satellite (see Thing 8). This enables UK stations to exchange pictures with amateurs from Asia, through Europe and Africa and across to South America. Also, the International Space Station carried an ATV transmitter during Tim Peake's mission and UK amateurs were able to receive pictures directly from the ISS.

5: Beyond the Everyday

Equipment for ATV

ATV can use the same transverters, preamps, power amplifiers and antennas that would be used for voice and Morse on the UHF and microwave bands. The digital TV signals are received and decoded using a special satellite TV tuner on a home-constructed PCB with a USB interface (the MiniTiouner). The tuner covers 144-2450MHz so it can be used directly after a preamp on the lower bands, or after a down-converter (or satellite LNB) on the higher bands. The free control and display software is run on a Raspberry Pi or a Windows PC. Full details can be found on the British Amateur Television Club (BATC) Wiki at https://wiki.batc.org.uk/MiniTioune.

The transmitted TV signals can be generated using a Raspberry Pi or a PC and a software defined radio such as a Pluto or a LimeSDR. If using a Raspberry Pi with the open source Portsdown software, a Raspberry Pi Camera or specific webcams can be used. If using a PC then any compatible webcam is suitable. The Raspberry Pi-based Portsdown transceiver is a good starting point for beginners and provides the ideal basis for a portable ATV system. It is described on the BATC Wiki at https://wiki.

Photo 43.1. Portable ATV can cover a range of 150km, especially from hilltops using portable gear like G8GKQ's portable station. Photo: Dave Crump, G8GKQ.

Figure 43.1. Saro, VU3OBR received in the UK through the Es'hail2 satellite by Dave Crump, G8GKQ.

Figure 43.2. Amateur TV received directly from the International Space Station by Dave Crump, G8GKQ.

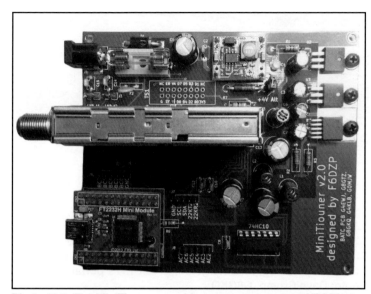

batc.org.uk/The_Portsdown_DATV_transceiver_system. Operation through the QO-100 satellite requires a modest, but neighbour-friendly, satellite dish using a standard commercial LNB for receive and a 13cm (2.4GHz) power amplifier for transmit.

The British Amateur Television Club Wiki is a great starting point for finding out about this exciting aspect of amateur radio – see https://wiki.batc.org.uk/. Indeed we thank BATC's Dave Crump, G8GKQ, for his contribution to this "thing".

Photo 43.2. The MiniTiouner provides an easy route into ATV. Photo: Dave Crump, G8GKQ.

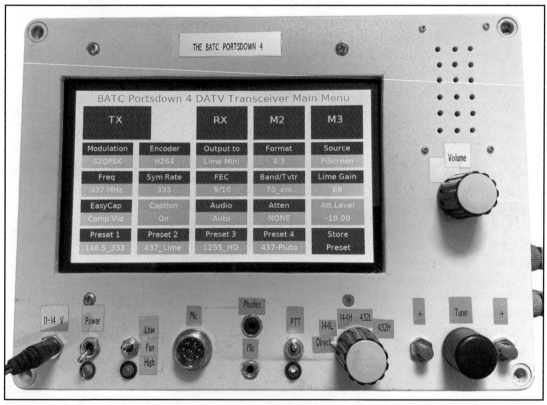

Photo 43.3. The Raspberry Pi-based Portsdown transceiver is a good starting point for beginners. Photo: Dave Crump, G8GKQ.

5: Beyond the Everyday

Thing 44 – Transmit from High-altitude Balloons

Dubbed 'the poor man's space programme', ARHAB, or Amateur Radio High-Altitude Ballooning, offers amateur experimenters the opportunity to launch balloons well into the stratosphere. More specifically, amateur balloons typically reach an altitude of 30-40km, which is well above the 9-12km cruising altitude of commercial jet aircraft. At this altitude, the temperatures are as low as -60°C and the air has only 1% of its density at ground level.

Typically, a flight will take one or two hours. At launch, the balloon will be around one or two metres in diameter and filled with helium. It will take an hour or two to reach its maximum altitude, at which point it will have expanded to a diameter of 7-11m before bursting. The payload will then parachute to the ground and will, hopefully, be retrieved by a team who has tracked its descent. Depending on the wind speed, the landing site might be as much as a few hundred kilometres from the launch site.

We should also mention pico balloons. They are much smaller than regular balloons and carry very small payloads, typically weighing just 10-20g. By using a small amount of helium, instead of bursting they float near the top of the troposphere, often for days or sometimes months.

The skills needed to build, launch and track a high-altitude balloon go beyond those that most radio amateurs can contribute, so there's plenty of opportunity for collaboration. However, our emphasis here is on the aspects of such a project that involve amateur radio and electronics. These systems will typically be used to transmit the location derived from on-board GPS, and periodic still photos. Some balloons also carry more specialised sensors and could return data such as temperature, atmospheric pressure, relative humidity and even levels of radiation.

There's an educational element to ARHAB too. Indeed, it's been suggested that it's a great way to inspire the next generation of scientists and engineers who, perhaps, will be part of the ever-growing space industry. In fact, amateurs have engaged young people in their projects in schools worldwide.

You should be aware that you shouldn't be tempted to fly a high-altitude balloon without adhering to the local regulations because of the risk to air traffic. Regulations differ in what sizes are regulated, but in the UK you need to apply for permission for the

Photo 44.1. The payload is suspended from the balloon by strong cords, sometime in several packages. Photo: StratoLab - Ricerche nella stratosfera,

Photo 44.2. Launching a high-altitude balloon will be the start of a two-hour, nerve-wracking flight. Photo: IBM Research.

launch of any high-altitude balloon.

Radio Equipment

Turning to radio regulations, the onboard transmitters have not, traditionally, been amateur radio equipment because airborne amateur operation was not allowed in the UK. Recent changes to the amateur radio licence have relaxed this somewhat, and we'll touch on this later. However, we'll concentrate here on the conventional solution which is likely to continue to be used in the near future. That solution is to transmit using licence-free modules that operate within the ISM band that covers 433.050-434.790MHz. These are low-power modules which, under normal conditions, provide a typical range of around 100m. While that would be totally inadequate for ARHAB, experimenters report ranges up to a few hundred kilometres because signals are not attenuated by proximity to the ground. Ranges are further improved by using better receivers and antennas than the cheap receiver modules that are typically used with 433MHz transmitter modules. On that point, you'll notice that the ISM band is within the 70cm (430MHz) amateur band, so using amateur equipment for reception is a possibility.

Although the British amateur radio licence now allows airborne operation, this only applies to those bands which are not shared by other services. So, because the 433MHz ISM band falls within the 70cm amateur band, this band cannot be used on high-altitude balloons, perhaps to permit higher-powered transmissions. However, the 2m (144MHz) amateur band is allocated exclusively to the amateur radio service, as are several HF bands. As such, it's possible that we might start to see true amateur operation from high-altitude balloons in the future.

Tracking Stations

Even if you choose not to become part of a project team to design, launch and track a high-altitude balloon, you can still get involved by tracking amateur balloons that others have launched. You might have to wait a long time for a balloon to fly over your house, though, so either you'll have to travel to the vicinity of balloon launch, as published in its flight plan, or use an online service. Commonly used flight tracker websites such as Flightradar24 aren't suitable because most

5: Beyond the Everyday

Figure 44.1. The Sondehub website allows the positions of amateur high-altitude balloons to be monitored via a network of volunteer trackers. Photo: StratoLab - Ricerche nella stratosfera.

balloons don't carry the ADS-B transmitters which allow aircraft to be tracked. However, the amateur community has provided similar functionality. So long as the flight has been registered with the site, the current location of balloons can be monitored at https://amateur.sondehub.org. Their locations are displayed on a map, together with information such as the horizontal and vertical velocity. Graphs can be generated for the reported altitude and any data provided by on-board sensors.

Because of the lack of ADS-B on amateur high-altitude balloons, the Sondehub website relies on reports uploaded by a network of amateur-operated tracking stations. Reportedly, these can be put together cheaply using low-cost SDR modules, so this could be your first step into 'the poor man's space programme'.

Photo 44.3. This photo, taken from an amateur radio high-altitude balloon clearly shows the blackness of space from its lofty location. Photo: K2GTX.

123

Thing 45 – Break a World Record

For most people reading this, achieving a world record for the triple jump or the 200m backstroke will be unlikely in the extreme. Don't give up on the idea of breaking a world record, though, since amateur radio might just offer that opportunity. To be fair, we'd have to suggest that you don't hold your breath, but we trust that this "thing" illustrates something of the amazing opportunities on offer.

Distance Records

Perhaps the most obvious records you might be able to break are those that relate to distances covered by transmissions. Needless to say, these will relate to particular bands or, perhaps, special conditions such as especially low-power operation. As such, they're unlikely to be reported in Guinness World Records. After all, these nuances will go over the head of most people who won't be au fait with the finer points of amateur radio. Even so, you could earn the recognition and admiration of your fellow radio amateurs.

You need to pick your band carefully though. For example, there are surely no distance-related world records still to be broken in most of the shortwave bands. After all, there's a limit to the distance between any two places on the Earth's surface, and it's probable that the maximum possible distance was covered many decades ago. This seems especially likely since several of the world's most populated cites have an antipodes on dry land. For example, Seoul in South Korea and Buenos Aires are 19,419km apart which is almost the maximum possible separation.

While you're unlikely to find references to distance records on the 20m or 15m amateur bands, though, the same isn't true of some of the other bands. As we saw in Thing 5, amateur operation at VLF and below is comparatively new and records are being broken on a regular basis. From a few kilometres just a decade or so ago, contacts spanning the Atlantic have now taken place in the so-called dreamers' band, that's the region below 9kHz. Meanwhile, at a higher VLF frequency of 17.47kHz, contact has been made between Germany and Tasmania, Australia, a distance of 16,805km. Other bands in which records will probably continue to see records broken are in some of the microwave bands (Thing 6), and via beams of light (Thing 46). This latter area, while not strictly speaking taking place in an amateur band, is a pursuit commonly

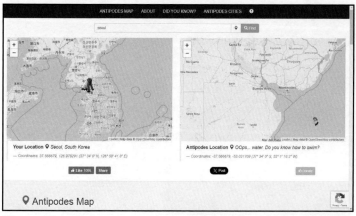

Figure 45.1. Distance records on the shortwave bands probably can't be beaten because there's a limit to how far any place can be from its antipodes. At VLF, though, these limits are yet to be reached.

undertaken by amateurs.

Guinness World Records

If you really do fancy the idea of seeing your name in lights – or Guinness World Records, actually – don't give up all hope just yet. In 2004, Guinness World Records Ltd recognised an amateur radio achievement, possibly for the first time, in awarding a certificate for the high-speed telegraphy achievement of Andrei Bindasov, EU7KI. According to the Guinness database listing, "On 6 May 2003 Andrei Bindasov (Belarus) transmitted 216 marks of mixed text per minute during the 5th International Amateur Radio Union World Championship in High Speed Telegraphy in Belarus."

And this achievement wasn't a one-off. Just a year later, the organisation awarded a certificate to Finnish radio amateur Jukka Heikinheimo, OH2BR for the record number of contacts made by an individual from one location in one year. Operating as VP6BR from Pitcairn Island, Heikinheimo made 56,239 contacts between January 25 and April 21, 2000. This record has been broken several times more recently. Since 2018, the record for most contacts in a year has been held by Austrian radio amateur Andrey Fedorov, who clocked up 93,856 contacts while operating, as 4U1A from the International United Nations Amateur Radio Contest DX Club (ARCDXC). By our calculations, that comes to a mere 257 contacts per day, so what are you waiting for?

Photo 45.1. Morse code might be considered archaic – at least outside the amateur radio community – but it's allowed some amateurs to break a world record.

Photo 45.2. If you fancy a certificate from Guinness World Records, or even an entry in the annual book, amateur radio might just provide that opportunity. Photo: MrBadawi.

Thing 46 – Talk via a Beam of Light

Given that this book is entitled 50 Things you can do with Amateur Radio, you might be surprised at this reference to light-beam communication. It's not as if radio waves and light beams are one and the same thing, after all. Both are forms of electromagnetic radiation, but that's about as far is it goes. Even so, because radio amateurs are often up for a challenge, and light-beam communication certainly offers this, it's really not too surprising that much of the work here has been undertaken by radio amateurs.

As a bit of background, the highest frequency band in the radio spectrum is the THF (Tremendously High Frequency) band which covers 300GHz to 3THz, and doesn't appear in Figure 3.2 because it doesn't contain any amateur bands, except by special permission. Above 3THz we find infrared radiation, which continues up to 430THz, and then there's visible radiation, i.e. light, which tops out at 750THz. In terms of wavelength, the visible spectrum runs from 700nm at the low-frequency end, i.e. red light, to 400nm for violet light.

We already know something about operation in the amateur microwave bands (Thing 6) and, in particular, that it's pretty much limited to line of sight and that expected ranges are short. The highest frequency regular amateur band is the 1mm band (241GHz) and the lowest frequency end of the visible spectrum is more than three orders of magnitude higher. We might expect, therefore, that the challenges in communicating at the top end of the radio spectrum will be exacerbated in the visible spectrum, and that's true in most respects. However, it doesn't apply to all aspects of communication by light, as we're about to see.

Photo 46.1. An LED, resistor, battery and Morse key are all you need to start with light-beam communication, but you're going to need more sophisticated gear to transmit further than a kilometre or so.

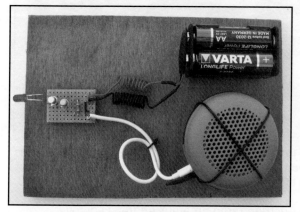

Photo 46.2. Even transmitting a speech signal via light is trivially simple as you can see from this receiver, which uses an LED as a detector. Expect ranges to top out at a few metres though.

Simple Experiments

Generating and receiving a signal at the top end of the microwave spectrum is tricky, and that's an understatement. Plug-and-play black box amateur rigs top out with the 23cm (1.2GHz) band so, for a start, would-be operators of the higher bands will need to break out the soldering iron. And that's not just for attaching components to PCBs – some aspects of the highest amateur bands seem to have more in common with plumbing than electronics. Not only that, but transmitters are grossly inefficient. All this changes with light, after all, light bulbs aren't especially expensive and LEDs are quite efficient.

So, if you want to experiment with light-beam communication, nothing could be easier. An ordinary red LED, resistor, battery and Morse key are all you need to transmit Morse code over 100m or so, for reception by eye. And by choosing a more suitable LED, perhaps higher powered and with a narrower beam, a kilometre or more could be achievable. Alternatively, if you fancy transmitting an audio signal, you could do that with very simple circuitry, albeit over a much shorter range. Going beyond these simple experiments, though, requires much more ingenuity. It's here, therefore, that we turn our attention to those amateurs who have set themselves the challenge of transmitting over almost unbelievable distances.

Photo 46.3. Barry Chambers, G8AGN/P in contact with G0EWN/P over a 112km path, using his high-power, LED-based optical equipment. Photo Barry Chambers.

Ultimate Achievements

Unlike most radio communications, communication by light can be hindered by the troposphere, which is the lowest layer of the atmosphere. This is where most of the air and water vapour in the atmosphere is found, so it's also the layer affected by weather. It'll come as no surprise, therefore, that amateur experiments have to pick favourable meteorological conditions to attempt long-distance communication. In a nutshell, we're talking of clear atmospheric conditions.

Using high-power LEDs, and accurately aligning the transmitter and receiver has allowed amateurs to communicate via Morse code over more than 100km, as you can see in Photo 46.3. However, because communication by light is a line-of-sight phenomenon, situating the transmitter and receiver on high ground offers a major advantage. Even so, there's a limit to just what's achievable.

However, the term 'line of sight' can be rather more nuanced, and this can give rise to communication over the horizon. Here, water vapour in the troposphere is actually an advantage rather than a hindrance. The secret is to fire the transmitted beam into the air rather than along the surface. This sounds paradoxical, certainly, but it can allow the signal to be received over the horizon, having being reflected from clouds between the two stations. Believe it or not, this allowed Australian radio amateurs to make contact between Tasmania and the mainland, a distance of 288km.

Thing 47 – Bounce Signals off Meteor Trails

Back in Thing 12 we looked at propagation, that's the means by which signals get from A to B. In the main, this related to the shortwave bands where radio waves can propagate around the globe by reflecting back to Earth from the ionosphere. We only made passing reference to VHF or UHF propagation, though, because mechanisms that allow communication much beyond line-of-sight are much rarer. In particular, we only mentioned Sporadic-E propagation, which works in a similar way to shortwave propagation, but is much rarer and doesn't offer worldwide communication. Here, and in the next "thing", we look at two other methods that work in the VHF band, and one that also works at even higher frequencies. Being in this chapter, though, they aren't exactly everyday.

Meteoroids and Meteors

According to the International Astronomical Union, a meteoroid is "a solid object moving in interplanetary space, of a size considerably smaller than an asteroid and considerably larger than an atom." In particular, a meteoroid is between ten micrometres and a metre in diameter. Some of these meteoroids enter the Earth's atmosphere – and are then referred to as meteors, or colloquially a shooting star – burning up to leave a visible trail. Most of these meteors are only the size of a grain of sand, but their trails can be a metre wide and several kilometres long.

Millions of meteors fall through the atmosphere every day, although there are more in the summer and during the few hours after dawn. Over and above these general trends, there are times of heightened meteor activity which occur when the Earth passes through debris left behind by comets. These so-called meteor showers show a peak in activity lasting a couple of days, the most important ones being the Perseids in mid-August and the Geminids in mid-December. At their peaks, these showers can put on a magnificent show, with up to 150 per hour being visible from some locations.

Oh, and in case you're wondering, meteorites are those meteors that get through the atmosphere partially intact and fall to Earth.

Photo 47.1. Meteors, or shooting stars, might be an impressive spectacle but they also permit long-distance VHF communication. Photo: Navicore.

Meteor Scatter

While the most obvious evidence of meteors are their visible trails, their passage through the atmosphere also causes the ionisation of atmospheric gases at an altitude of about 70-110km. As you'll recall from our discussion of propagation, radio waves can be reflected by ionised particles in the atmosphere and this can give rise to long-distance communication. While this is most commonly associated with particles ionised by solar radiation, the same is true of meteoritic ionisation. Where it differs, though, is in the frequency affected. For while solar ionisation mainly affects the shortwave bands, ionisation caused by meteors has an impact on VHF propagation.

This method of communication is commonly referred to as meteor scatter. Because the ionisation is at a lower altitude than that caused by solar ionisation, the communication range isn't as great. However, ranges significantly greater than achievable under normal conditions apply, with ranges up to around 2,000km being possible. Most amateur operation is on the 6m (50MHz) band, but, because the improvement in range is more spectacular there, the 2m (144MHz) band is also popular, albeit more challenging. Reportedly, the 70cm (430MHz) band has also been used, but it's very unlikely in the extreme that you'll have much success in this band.

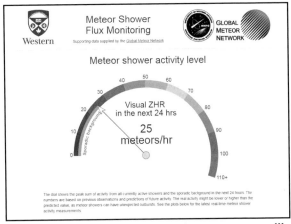

Figure 47.1. Dates of meteor showers are readily available, but at other times sites showing meteor activity, like this one at https://globalmeteornetwork.org/flux/ can be useful.

Practically, a reasonably high-gain antenna is needed. However, a compromise is required because a high gain will result in a narrow beam width, so fewer meteor trails will be reachable. However, a much more fundamental issue has to be addressed. The ionised trails on which propagation relies don't last long. While they can be useable for a few seconds, a quarter of a second is considered more typical. And this has an impact on the transmission modes that are suitable. Traditionally, Morse code has been used, sent automatically at a speed of several hundred words per minute. Even so, it would often be necessary to employ the trails of several separate meteors, so a contact could take several minutes. Today, however, high-performance data modes tend to be used. Popular here are FSK441 and MSK144 which share many features of other weak signal data modes (see Thing 11), but are tailored specifically for meteor scatter. To cut a long story short, these modes repeat short messages multiple times with the aim of assembling the full message from snippets received from several meteor trails. And here we feel the need to reiterate the well known saying that patience is a virtue.

Photo 47.2. Don't expect to see as many meteors as this long exposure photo might suggest, but you'll probably need to use several successive meteor trails to make a single contact.

Thing 48 – Bounce Signals off the Moon

Amateur radio via meteor scatter is pretty amazing and, to the uninitiated, quite unexpected. And here we're looking at another way of increasing the range of radio in the VHF band, but also in the higher UHF and microwave bands. Called moon bounce communication – or EME (Earth-Moon-Earth) to its adherents – it does exactly what the name suggests. It allows you to increase your range substantially, almost to the opposite side of the globe, and it surely takes some beating for 'wow factor'.

Lunar Facts

It might allow the otherwise impossible to be achieved, but using the Moon as a reflector is by no means trivial as a few facts and figures will reveal. The Moon is 3,474km in diameter, and it orbits the Earth at an average distance of 384,400km. The upshot of this is that the Moon has an angular size of half

Photo 48.1. It might be a familiar sight but that doesn't make bouncing radio signals off the moon any less amazing.

a degree at the Earth's surface, so it occupies only about 0.001% of the area of the sky. Of course, pretty much all antennas in the bands used for EME have a reasonably high gain and, therefore, they're not unidirectional. However, for normal use on 2m (144MHz) an antenna's beam width is unlikely to be much less than 40°. Needless to say, if you use this sort of antenna, very little of your signal would reach the Moon. And it gets worse because the apparent size of the Earth from the Moon is similarly tiny, but that's not all. Since the signal reflected from the Moon won't benefit from a directional antenna, the percentage of the return signal reaching the Earth will be even lower. Add to that the fact that the Moon reflects only about 10% of a radio signal reaching it, and we can see that signal levels are going to be minute.

The size of the target and the Moon's limited reflectivity are only part of the story. As we'll see later, modern weak signal modes are now important tools in EME communication. Because these use much narrower bandwidths than speech or Morse code, accurate tuning is needed, and this brings us to the next challenge. Due to the rotation of the Earth, the Moon is in motion relative to the surface of the Earth, and this causes the frequency of the received signal to vary due to the Doppler effect. Continually tracking the received signal is, therefore, essential.

We're not going to get embroiled in the details here, but this is by no means the end of the ways in which a Moon-bounced signal departs from what we might hope and expect. However, we will refer to one other phenomenon, which is actually more of a benefit than a hindrance. That 384,400km distance to the Moon means that the round-trip time is 768,800km, and this results in a delay of around two and a half seconds between a signal being transmitted and it arriving back on Earth. This, therefore, provides an opportunity for you to check, yourself, whether your signal is successfully traversing the path from the Earth to the Moon and back.

Practical Considerations

At one time, the huge path loss that we looked at in the previous section had some serious consequences. In particular, high transmitter powers were used, and high-gain antennas were also needed. Indeed, the antenna shown in Photo 48.2 – comprising eight phased 16-element beams – would have been not at all unusual. And while a 16-element beam for the 2m (144MHz) band would be considered a lot better than average for everyday use, this phased array is undoubtedly huge. What's more, antennas for higher frequency bands tended to have an even higher gain due, in part, to the higher path loss at these frequencies. If you want to be able to communicate by speech then large antennas and high powers are still necessary but, according to EME enthusiasts, there are better ways, as we'll see next.

Like so many other areas of amateur radio, weak signal modes like those we looked at in Thing 11 have revolutionised Moon bounce communication. Their use has allowed amateurs with much more modest hardware to succeed in this most unusual form of amateur radio operation. In fact, it's been suggested that an 80W 70cm setup using about a 12-15 dBi Yagi (that's perhaps a 15-element beam which is only about 2m long at this frequency) works well using digital modes like the JT65. Given that EME involves transmitting over a distance equivalent to over 19 times around the Earth, this surely isn't bad going.

Photo 48.2. For conventional modes, a huge antenna was an essential part of any Moon bounce setup, like this one used by EA6VQ.

Photo 48.3. I2FZX's 6m parabolic dish for the UHF bands isn't at all unusual in the upper echelons of EME communication.

Thing 49 – Talk with Astronauts

Our next "thing", while not involving bouncing signals off anything, continues in a space theme, and it's no less surprising. We're talking of chatting with astronauts via amateur radio, and this is no longer something only for space agencies like NASA. To be frank, you probably won't end up talking to astronauts yourself, after all they'll be very much in demand, but that doesn't mean you can't get involved.

Almost certainly, any astronauts who are operating an amateur radio will be on board the International Space Station (ISS). Astronauts on other missions will probably have their hands full, but astronauts on the ISS will usually be in the space station for several months, so they'll definitely have some down time. Several ISS astronauts from a variety of countries have held amateur radio licences, and between them have contacted thousands of amateurs.

Photo 49.1. The International Space Station might be an unusual place for an amateur radio station, but there are commonly amateurs on board.

We're not providing much in the way of technical advice here, but information about which frequencies to listen or and knowing when to find astronauts on the amateur bands is plentiful. Instead, our aim is to illustrate what's possible and hopefully motivate you to take the next step.

Listen to Astronauts

Listening to a much sought after rare station is always easier than actually talking to that station. After all, lots of people can listen at the same time while only one can have a two-way contact. The limit to the number of people who can talk to rare stations is further exacerbated by the fact that lots of amateurs will almost certainly be trying. And stronger stations will probably prevent you from being heard. So, our suggestion is that you start out by listening to astronauts. This will be a rewarding opportunity in itself, but it could also make you better placed to take the next step. If you have perseverance, that next step might just be talking to an astronaut yourself or, alternatively, you might fancy the idea of helping young people do just that, which is our next topic.

Encourage School Students

In Thing 41 we investigated how you could play a part in giving people a passion for radio. More specifically we saw how young people could be drawn into amateur radio and, in so doing, be given a passion for science, technology, engineering and maths. The ARISS Programme (Amateur Radio on the International Space Station) has just this aim, by offering schools the opportunity for their students to talk to astronauts. As ARISS says, "ARISS lets students worldwide experience the excitement of talking directly with crew members of the International Space Station, inspiring them to pursue interests in careers in science, technology, engineering and math, and engaging them with radio science technology through amateur radio."

These contacts don't happen on an ad hoc basis. Instead there's a formal application process that's open to formal and informal education institutions and organisations worldwide. Needless to say, applications exceed the available opportunities but, even so, over a million people are reported to have been involved in one way or another. Of relevance to our discussion here, many of those people are radio amateurs who support the host educational organisations. To get a feel for these events, we quote from a news story published by St. Peter's CE Primary School in Broadstairs, Kent.

Photo 49.2. NASA astronaut Doug Wheelock uses an amateur radio station in the Zvezda Service Module of the ISS.

"Space and science fans at St Peter's Juniors in Broadstairs have chatted live with an astronaut on board the international space station via a radio link as it soared 1,430 miles above. Children quizzed Jasmin Moghbeli as she sped through the cosmos at 17,500 mph via a link made possible by ARISS, the Amateur Radio on the International Space Station team, and Hilderstone Radio Society... Weeks of waiting distilled to the historic moment when pupil Isabella read out the official call signal to the space station and Jasmin's reply crackled through the airwaves to gasps, cheers and flag waving from the excited young audience in the hall... (Science teacher) Mr Williams described the event as 'a fantastic opportunity for our children and other pupils from visiting schools to share a ground-breaking moment with us.' He highlighted the significance of amateur radio to the project. He explained: 'It represents the very essence of human ingenuity and the unyielding desire to connect, explore, and communicate. It has been the bridge that has connected our school to the cosmos, enabling our students to engage in conversations that transcend the boundaries of our planet.' 'Through amateur radio, our children have delved into the intricate world of frequencies and signals, learning not only about the wonders of space but also the importance of effective communication.'" Perhaps you too could play your part by helping to facilitate such a worthwhile event.

Photo 49.3. A question-and-answer session between students from the Goddard Child Development Center, local students and residents aboard the International Space Station.

Thing 50 – Contribute to the State-of-the-art

If you think that all the major developments in radio communication have been made by academics and engineers in industry, think again. Radio amateurs have an enviable heritage in this area, and continue to push the state-of-the-art to the current day.

A History Lesson

Before turning our attention to today's contributions to the state-of-the-art, it will be informative to look back over the decades. As a word of introduction, though, we need to say a few words about why we've included some pioneering individuals who might not seem to qualify. That's because they didn't all hold amateur radio licences. However, in the early days, the use of radio wasn't regulated so there was no concept of licensing people to transmit. So, we're using the word 'amateur' in its strictest sense. Derived from the Latin word amator, the original meaning of 'amateur' is someone who engages in something for the love of it. Such a passion didn't mean these individuals couldn't go on to use their expertise in industry or academia, though, indeed several did exactly that.

Whether or not some of the early pioneers – and here we could mention Samuel Morse – could be considered as amateurs is questionable since it rather seems that many were purely commercially motivated. Turning to Marconi, though, his status as an amateur experimenter is more debatable. He is described as conducting his first experiments at the family home at the age of 20, two years before making a patent application and four years before setting up the Marconi Company. So, just possibly, we might be able to call the Father of Radio, Guglielmo Marconi, the first radio amateur. What greater contribution could a radio amateur make?

A name that is more commonly cited as a hugely-influential radio amateur, in these pre-licence days, is that of Edwin Howard Armstrong. And while he went on to sell one of his patents to RCA in 1922 for $200,000 (today's equivalent of $3,671,631) and 60,000 shares, his fascination with radio as a boy is well documented. Indeed, in 1904 at age 14, motivated by Marconi's achievements, he started experimenting with radio equipment in the attic of his family's house. Dubbed "radio boy", reportedly he constructed a makeshift backyard antenna tower that included a winch for hoisting himself up and down its length, much

Photo 50.1 His status as an amateur might be questionable, but Guglielmo Marconi's contribution to the state-of-the-art is unquestionable.

to the consternation of his neighbours.

But his technical contributions were hugely influential. Here we must mention the regenerative and later the superheterodyne receiver, concepts that will be familiar to pretty much anyone who is au fait with radio receiver circuits. And although the SDR (software defined radio) is starting to sideline the superheterodyne architecture, it's still going strong, over a hundred years since Armstrong first demonstrated the concept. Arguably, though, Armstrong's next invention was even more influential. We're thinking here of FM

Photo 50.2. Inventor of the superheterodyne receiver, Edwin Howard Armstrong, describes the concept in New York, in 1922.

Photo 50.3. This AT&T Beverage antenna, comprised a pair of straight wires several kilometres long grounded through a resistor at the end, provided the directionality needed for a transatlantic link.

Photo 50.4. In academia, Joe Taylor is best known for discovering a first binary pulsar by radio astronomy. But radio amateurs associate him with his low power data modes that also rely on signal processing principles. Photo: Mariordo (Mario Roberto Durán Ortiz).

(frequency modulation) that has been used for broadcast radio for almost a century, and is widely used today by radio amateurs, especially at VHF and UHF.

We could also mention radio amateur Harold Beverage, W2BIE, who invented the Beverage antenna in 1922, and the 1930s development of the W8JK antenna by John Kraus, W8JK. Both developments were considered innovative and continue to be used by radio amateurs to the present day. The Beverage antenna also enjoyed commercial and military success. It was used, for example, by AT&T at their longwave receiver site in Houlton, Maine, to provide a transatlantic telephone service around 1927.

Recent Pioneers

Ground-breaking developments by radio amateurs weren't just in that long-ago golden age of radio. In the 2001, physics Nobel laureate Joe Taylor, K1JT developed state-of-the-art methods of digital communication intended for use over marginal communication channels. As such, they allowed contacts to be made with low power or over difficult paths, where communication via speech or Morse code would be impossible. Several such weak signal modes have been developed by Joe over the years (see, for example JT65 and WSPR in Thing 11), each aimed at different applications and/or different frequency bands. These modes have become hugely popular among radio amateurs, and have found application beyond the amateur community. Indeed, Joe routinely acknowledges the prominent role which his amateur radio background has played in his professional and academic success, and commonly refers to the important contributions still to be made by dedicated amateurs in the fields of radio astronomy and SETI (search for extra-terrestrial intelligence).

Talk of Nobel prize winning radio amateurs might beg the question of whether 'ordinary' radio amateurs can make a contribution to the state-of-the-art. Well, it just so happens that the answer is yes, as I've learned myself. My main passion in amateur radio is sub-surface communication, as used by cavers and cave rescue teams, a subject you can read about in Thing 36. This led me to undertake work on measuring the attenuation of microwave signals along a variety of cave passages and developing a variety of methods to predict propagation using software modelling. Several of these developments have been published as papers in academic journals, a requirement for which is a successful peer review. The peer review process confirms technical accuracy, certainly, but it also recognises the work as being novel or, if you like, state-of-the-art. Perhaps if, like me, you consider yourself an 'ordinary' amateur, but you have an interest in something a bit out of the ordinary, we'd encourage you to pursue it. And if you're prepared to commit some time and effort into your project, you too might be able to make a contribution to the state-of-the-art.

6

Next Steps

So, there you have it - 50 Things you can do with Amateur Radio. And the chances are that quite a few of them were unexpected, given the huge diversity on offer in amateur radio. Expected or not, though, we trust, should you take the next step, that there's plenty in this multi-faceted hobby to keep you occupied, enthralled, entertained and educated for quite some time. So, what is that next step? How can you get started in amateur radio or, if you've previously engaged in amateur radio, how can you return to your former passion?

Existing Radio Amateurs

First of all, because this is the easiest case, here are a few comments for the radio amateur who has previously enjoyed this hobby, but has taken a break. Much of what you previously enjoyed will still apply, but we trust that we've introduced you to some new and exciting aspects of the hobby. After all, amateur radio doesn't stand still, and perhaps this is what has caused you to consider returning to the fray.

The good news is that you'll probably still be able to use your amateur radio licence if you live in the UK, Guernsey, Jersey or the Isle of Man. Even if you've not been operating on the amateur bands for some time, your licence will still be valid so long as you've continued to validate it

6: Next Steps

on the Ofcom website at least every five years. If your licence has lapsed, though, it's a bit more difficult but, you should be able to reinstate it. This involves applying to Ofcom for a reinstatement with proof that you're entitled to hold a licence and that you previously held the callsign in your own name.

Moving on, probably the best bit of advice we can offer is to get back into contact with nearby amateurs, even before you start to assemble your station again. As you're no doubt aware, making contact with a local club is, perhaps, the best way of doing that. And since enthusiasm is infectious, this will probably give you the added incentive to get back on the air. What's more, it'll probably give you the opportunity to learn about some of the new aspects of the hobby that have come to the fore since your last foray into amateur radio. Beyond that, it's really up to you. After all, you've been this way before, so it'll surely come naturally.

Newcomers

Next we come to those who haven't previously dabbled in amateur radio but who've been motivated to learn more and, hopefully, to obtain a licence. The biggest step is to get that all important licence and we'll come to that in due course. However, there are things you can do in parallel, and these things are also a good idea, even if you've not yet decided to obtain a licence.

We've already looked at both these steps, but they're both well worth reiterating here, albeit briefly. First of all, as we saw in Thing 1, you can make a start in amateur radio even without a licence, indeed this is commonly a first step into the hobby. That first step is to get hold of a suitable receiver and start listening on the amateur bands. In fact, you don't even need to buy a receiver as there's an even simpler solution in the form of a WebSDR. It might not give you quite the same technical experience, but it offers a no-cost, no-hassle way of gaining your first experience of amateur radio operation.

The next bit of advice is exactly the same as we gave to the lapsed amateur who is keen on returning to the hobby. That advice is to visit a local amateur radio club and, ideally, become a member – our topic in Thing 38. You'll be encouraged and motivated by other amateurs, you'll be able to learn about so many aspects of amateur radio, and you might be given the opportunity to see amateur radio operation at first hand. And finally – and this leads into our next subject – your local club might offer training to help you obtain a licence.

Obtaining a Licence

Wherever in the world you live, you'll only be able to transmit on the amateur bands if you hold a licence. If you live in the UK, Guernsey, Jersey or the Isle of Man, here's what you need to know, although if live elsewhere you'll have to delve into the details yourself.

There are three levels on licence – foundation, intermediate and full – and you're give more privileges as you move from one class to the next. For most people starting out in the hobby, the first step is to qualify for a foundation licence. So, while you'll probably want to obtain a full licence in due course, here we're just looking at how to get a foundation licence.

You're entitled to apply for a licence only after you've taken and passed an examination for a particular class of licence. This requirement is to ensure that you're able to operate a station safely and without causing interference to other users of the radio spectrum. It also proves that you understand the conditions that will apply to your licence.

The foundation level syllabus covers the following main subjects:
- Licensing conditions and station identification,
- Technical aspects,
- Transmitters and receivers,
- Feeders and antennas,
- Propagation,
- Electro magnetic compatibility (EMC),
- Operating practices and procedures,
- Safety.

It's considered that it takes 10 to 12 hours of study to adequately prepare to take an examination. However, some of this material will be second nature if you're technically-minded, especially if you have some experience of electronics. You can study for the exam by yourself using a book such as the RSGB book The Foundation Licence Manual for Radio Amateurs by Alan Betts, G0HIQ, via an online course, or by taking an in-person course organised by an amateur radio club.

Once you're up to speed on the syllabus you can book an examination and there are three options. You can take the examination online or on paper at an RSGB registered venue, usually a club, or at home online with a remote invigilator. The exam should be booked on the RSGB website and a fee of £35.50 is payable (as of 2024). The exam involves answering 26 multiple choice questions which you have one hour to answer. The pass mark is 19, which equates to 73%. If you opt for the online option, you will be given your result immediately after you complete the exam, together with detailed feedback about your performance.

Once you've passed the examination you can apply for a licence and, so long as you do this online via the Ofcom website, no fee is required. We look forward to meeting you on the amateur bands sometime soon.